TELECOMMUNICATIONS

*A Bridge to the
21st Century*

Edited by

DR. MEHEROO JUSSAWALLA

East-West Center
1777 East-West Road
Honolulu, Hawaii 96848

TELECOMMUNICATIONS

A Bridge to the 21st Century

1995

ELSEVIER

Amsterdam • Lausanne • New York
Oxford • Shannon • Tokyo

384
T2677

ELSEVIER SCIENCE B.V.
Sara Burgerhartstraat 25
P.O. Box 211, 1000 AE Amsterdam, The Netherlands

ISBN: 0 444 82325 5

DEDICATION

My efforts as the Editor and my part of this volume are dedicated to my grandson, Hormuzdiyar Henry Dasenbrock, whose interest in E-mail and Internet, at age eleven, is extraordinary. In the twenty-first century, Hormuzdiyar, when you will be speeding along the Global Infobahn, remember the precepts of the prophet Zoroaster: Manashni (good thought), Gavashni (good word), and Kunashni (good deeds).

CONTENTS

viii

ACKNOWLEDGEMENTS

I would like to acknowledge the cooperation of all my colleagues who contributed their time and expertise in preparing the chapters for this work. Laura Miho first processed many of the chapters on her computer while the dedicated and untiring efforts of Vivian Kiyonaga made the camera-ready version possible. My sincere thanks are due to Dr. Janis Togashi for her expert and patient editorial supervision of the manuscript.

INTRODUCTION

MEHEROO JUSSAWALLA

Societies around the world are today convinced that telecommunications technology will help to bridge our passage to the twenty-first century. Willy nilly, developed and developing countries are being driven by the dynamic changes that are being ushered into people's lives, their work habits, and their modes of socializing and interacting with each other in what is now recognized as the "Information Age." As innovative technology enters the markets, it creates a greater demand from users for acquiring it and drives the engine of growth towards speedier production, consumption, and exchange of information.

The United States of America has been in the vanguard of these trends as it plans to build a National Information Superhighway. Vice President Albert Gore has been the guiding light in this entire effort to build not just a National Information Infrastructure (NII) but a Global Information Infrastructure (GII) as well. The NII is to be constructed by the private sector in America, but the GII has so captivated the leaders of world countries that most governments are appointing commissions to examine its possibilities, its application to their countries, and how best each of them can get on the ramp of the global infobahn.

We are all aware that information is already moving across continents in real time and with the speed of light. Fiberoptic cables, whether submarine or terrestrial, powered by asynchronous transfer mode (ATM) switches are carrying enormous quantities of information in nanoseconds. The digitization of these networks and their convergence with satellites, geosynchronous or mobile, and with computers that are equipped with artificial intelligence are all creating electronic highways that encircle the globe at constantly declining costs. By the dawn of the next century, consumers will have a choice from among a flood of new devices and services that will integrate wireless voice, data, and compressed video images and deliver them worldwide to offices and homes using personal communications systems. To quote the Vice President of the United States, "telecommunications is an essential component of political, economic, social, and cultural development. It fuels the global information society and economy which is rapidly transforming local, national, and international societies and, despite physical boundaries, is promoting better understanding between peoples" (*Washington Times*, October 16, 1994). However, the gap between rich and poor nations is widening rather than narrowing because, in reality, technology although advancing is not yet affordable by people in rural and remote areas of developing countries. As such, those who control the switches of information and those who own these switches can transform the future of mankind for tremendous good or may expand the existing gap within a country or between countries. The concept of the information superhighway has gained the attention of the media in most countries and has penetrated all sections of society, from policymakers to users of cyberspace via the Internet.

Yet as Leonard Marks, the Advisor to the GII Commission in Washington, D.C. warned, "there are detours along the Information Highway" (*I-ways* 1995). He rightly points out that while the affluent countries including Europe, Japan, America, and Canada enjoy such services as direct dialing, receipt of faxes, and electronic interchange of data, and receive television programs from remote regions and are able hook up to the Internet, people of low-income countries in remote and rural regions yearn for dial tone. Two-thirds of the world's population have no access to plain old telephones (POTs). Large populations in countries such as India, China, and Africa live with less than one or two telephones per one hundred people. Low-income countries have 55 percent of the world's population but only 5 percent of the world's telephones. The World Bank estimates that $40 billion a year will be needed until 1999 to provide a telephone within walking distance of every village in the world.

Despite these gloomy statistics, China has plans to invest $100 billion in its telecommunications infrastructure by the year 2000 to give eight telephones per one hundred people and plans to open its markets to foreign participation in building its information industry. India, after decades of denying itself foreign technologies, succumbed to the call of global markets and is privatizing its telecommunications monopoly. The Southeast Asian countries of Singapore, Malaysia, Thailand, and Indonesia are catching up with Japan in providing their citizens with the most sophisticated technologies of computer networks, satellite-based linkages, and cellular telephony. All of these countries including China are publicizing their intentions for building a global information superhighway.

As each nation develops its own NII, it is necessary to understand how different NIIs will establish an integration of networks and how such integration will bring about liberalization of the international trade regime for information services. Furthermore, standards must be developed by both regional and international organizations to promote the interoperability of the networks. At the same time, it is necessary to examine the political, economic, and cultural impacts of a global superhighway. It becomes imperative to expand upon theories of pricing information and usage of networks. Despite all the high-speed networks and powerful PCs, the uncharted expanse of cyberspace known as the Internet is still not a safe and hospitable environment for businesses and consumers (Mitchell 1995). It is expected that the enabling software will change this and encryption on the World Wide Web will prevent misuse of the system. This has led to a great cyberspace software race which will create an electronic marketplace for the virtual corporation and communities to thrive without losing their digital valuables. These are some of the scenarios facing us in the twenty-first century and which are covered in this volume.

The groundwork for the explosion in growth was sparked by the global trend in deregulation started by the divestiture of AT&T in 1984. Monopolies around the world are being challenged and it is critical to understand what the outcome of this challenge will be in the future. In the first chapter, Eli Noam argues that the need to integrate different parts of the vast international information infrastructure will stand at the forefront of the telecommunications agenda in the next few years.

Noam observes that integration can be accomplished through three possible agents: end users, carriers, and systems integrators. Integration by the last of these agents provides several key advantages over the other two. Systems integrators can save end users from having to acquire the expertise required to bring about systems integration on their own. Furthermore, systems integrators, unlike major carriers, are not obliged to recoup billions of dollars in investments in specific technologies or networks. As a result, they can most effectively customize packages that best suit the needs of their customers. Consequently, Noam predicts that systems integrators will emerge as a key intermediary agent between users and carriers in the next decade.

With the rise of systems integrators, it is necessary to analyze what role, if any, government should play in regulating the new telecommunications environment. Despite deregulation, Noam cautions that government can still play a productive role in several areas. In particular, the government must help set standardization guidelines to ensure the interconnectivity of networks. The government can also play an important role in advancing the democratic structure of the information superhighway by ensuring universal services, affordable rates, nondiscriminatory access, and nondiscriminatory neutral use. As Noam warns, "unpopular" forms of communication on networks may be forced out unless the government intervenes to ensure a free flow of information. For example, in the newsgroup section of the Internet, known as Usenet, the Exon Bill may oust certain forums from having their own discussions. Additionally, public telecommunications operators (PTOs) and the governments of the United States and Japan, in particular, can play key roles in setting up access, lease, and interconnection arrangements that will allow systems integrators to operate without discrimination in a global free market. Thus far, it appears as if the Clinton administration has heeded the advice of Noam and others as outlined in the administration's intention to encourage private investment. In an address to the Center for Media Education in 1994, Vice President Albert Gore promised "to provide and protect competition, provide open access to the network, take action to avoid creating a society of information haves and have-nots and encourage flexible and responsible government action" with respect to the NII.

While the integration of the international information infrastructure advances the ability of the economies of the world to interact with each other, we still remain far from global economic and cultural integration. In Chapter II, Don Lamberton challenges the prevalent views of development and trade that have been hailed as the arrival of the global village. As Lamberton argues, in a world where the majority of the population does not reside within two walking hours of a telephone, the notion of the global village is "globaloney." Instead, he maintains that we are closer to regionalization than globalization.

To begin with, Lamberton subverts the globalist paradigm. He asserts that the vast asymmetries that continue to exist among nations in the areas of information and organizational systems render demands by economic integrationists for an internationally level playing field unreasonable. As he points out, the global economy is far from becoming integrated. Rather it consists of a set of individual economies with differing levels of adaptability, technological skills, and cultures. As a result, each economy has its own organization and information-using character.

Lamberton also criticizes mainstream economic theory for its assumptions about perfection and the myth of perfect information. He charges that mainstream economic theory incorrectly suggests optimizing at a point where no resources are allocated to information activities and at a time when in developed countries the most important sectors deal with information. He challenges the laissez faire policies advocated by neoclassical economists arguing that they ignored the powerful role that government played in the development of every industrial nation. Finally, he warns that globalists supporting economic integration disregard the effects that such policies may have on national goals of preventing profits from fleeing abroad. He emphasizes that restrictive and linked industrial structures, infrastructure disparities, and labor immobility violate the principal assumptions of economic liberalism.

While the term global village may be "globaloney," the fact remains that the advances in telecommunications have made the world a smaller place. One of these advances has been the development of cellular technology. In Chapter III, Yale Braunstein analyzes the growth of mobile and portable cellular technology in the Asia-Pacific region, which can be extended to other parts of the developing world. Braunstein begins with an examination of the current status of cellular penetration. Over the past ten years, cellular technology has vastly expanded the time and place where people can communicate. For example, users of mobile and portable services can remain in touch during the morning commute, while in isolated regions, or during emergency situations when a wireline is too distant to approach. While cellular services have expanded telecommunication opportunities for corporate users, smaller businesses, professionals, and emergency relief officials still find that there are problems restricting the expanded usage of the technology. Braunstein shows how existing analog cellular systems do not use bandwidth efficiently, especially as a result of the lack of standardization. Consequently, frequencies have become saturated. Privacy and health issues have reduced public enthusiasm to embrace the technology. However, future changes will address some of these difficulties and allow developing countries to exceed the "threshold" of 1.5–2 percent penetration, thereby ushering in the growth of mobile and portable systems. In the short term, the conversion to a digital standard will provide greater security for users concerned with privacy, will reduce inefficient use of the bandwidth, and will allow fax, data, and voice transmission. In addition to these advantages, mobile systems will be able to bypass local wireline carriers as they have already begun to do. In the long term, services will be extended significantly and hardware changes will enter the market. Braunstein identifies mobile and marine communications, trunked mobile dispatch, and cellular or satellite communications as the most likely forms that these new services will take. This has already occurred to some extent as we find low earth orbiting satellites being planned for China, India, and Latin America by Iridium, Globalstar, Geodesic, and Orbalcom, companies which claim to provide cost-effective cellular systems with roaming possibilities for the users. Already there is a scramble for the magnetic frequency spectrum in the United States for PCS bringing in billions of dollars in revenues for the Federal Communication Commission. In all, Braunstein concludes that the functionality gap between traditional services and cellular services will continue to exist, the economic and financial pressures of traditional wireline services will expand as cellular, satellite and other non-wireline services continue to bypass the traditional system to increase the choice for consumers.

One of the most publicized and exciting aspects of the telecommunications revolution has been the transition from multimedia to hypermedia, from non-interactive and linear technology to interactive and nonlinear technology. In Chapter IV, Dan Wedemeyer examines the growth of networking multimedia and hypermedia applications and analyzes the technical and legal dilemmas that such advancements pose. As Wedemeyer observes, human beings have always experienced the world through a variety of senses—sight, smell, sound, taste, and touch. However, technological limitations have restricted the telecommunications media from capturing the entire gamut of human sensation. For example, early movies could only capture a black-and-white moving image; traditional telephones could only transmit the sound of on-line parities. Increasingly, however, new advances are reducing the gap between telecommunications media and reality. The hypermedia and virtual reality networks of the future will make this possible.

However, every new innovation presents its share of problems. Technically, hypermedia networks will require a greater bandwidth than existing telephony systems can carry. Consequently some have advocated the adoption of Broadband ISDN. Wedemeyer suggests that hybrid wireless systems may better serve the development of visual databases by drawing on the narrow cellular telephone network for information requests and the broadband wireless cable channel for video requests. Legally, legislation continues to develop important protection of intellectual property rights in the area of hypermedia. But Wedemeyer warns that such rights should be neither overprotected nor underprotected. The 1995 legislation in the United States for telecommunications deregulation for both the telephone and cable industries has attempted to tackle the issues of such innovative technologies. In the short run, we can expect the decade of the 1990s to be "the visualization decade" in which technology will allow us to "see the sounds." In the long run, Wedemeyer cautions, the spread of information technology (IT) will precipitate broader global changes. Local cultures will be increasingly muted by the new global IT culture and the flow of information along the superhighway will dictate the flow of political power.

This relationship between telecommunications and culture is more closely examined by Syed Rahim in Chapter V which shows the ways in which this cultural structure is evolving. As he notes, culture represents the aggregation of both the shared and contested ideas, experiences, beliefs, and world views of all the different social, ethnic, political, and demographic groups within a society. National telecommunications systems incorporate elements of the culture which they represent. While national telecommunications cultures may differ significantly from one another in their approaches, they can retain similar core values. For example, while telecommunications policy in the United States has evolved from a series of value conflicts, telecommunications policy in Japan has emerged largely through a consensual process under the aegis of Japan Inc., i.e., the Ministry of Trade and Industry (MITI), the ruling Liberal Democratic Party, and Keidanran (the Federation of Economic Organizations). However, like the United States, Japan has embraced privatization and liberalization in the telecommunications sector. As Rahim argues, new cultural perceptions of the importance of telecommunications to post-industrial society paved the way for the vast structural changes that took place in the 1980s.

Both competition and deregulation are likely to impact culture in the future in a variety of ways. As Rahim notes, virtual classrooms and libraries will forever change the face of education; war simulations may change the way in which international relations are conducted in the future; the information superhighway will alter the formation, operation, and integration of social groups; and all the world will be a stage as satellite technology beams signals across the globe in real time. However, Rahim cautions that significant disparities still remain between developed and developing nations in the adoption of telecommunications technology. These disparities threaten to exacerbate the long-term division between the rich and poor nations of the world.

One of these impacts on society has been the increasing use of digital cash and electronic money. Paper-based cash as a medium of exchange has been under assault by electronic processing of checks, and credit and debit cards for years. However, cash has anonymity, an indelible quality that other forms of payment cannot match. As the use of electronic networks grows from Cirrus to the Internet, so do concerns about fraud and the privacy of transactions. New businesses are being designed to create a new kind of "electronic money" suitable for conducting cash-like transactions in the information age. Anthony Pennings in Chapter VI deals with questions about electronic money such as: Who will regulate digital dollars? How secure will they be? What will be the impact on financial markets and monetary policy? Pennings also discusses some of the issues that must be resolved if electronic money will compete with cash on a wide scale and what will be the impact of this revolution on information economies and cyberspace politics.

Many corporations today are developing their own forms of electronic cash known as E-cash. It is money that moves along multiple channels largely outside the established network of banks, checks, and paper currency overseen by the Federal Reserve in America and by the central banks in other countries. These players assume names like Digicash and CyberCash, and Citicorp is developing what it calls the electronic monetary system, which will be an infrastructure for using electronic money. Digital money is the ultimate medium of exchange for an increasingly wired world according to *Business Week* (1995, p. 66). Such digital money will allow users to transact business on-line. Pennings argues that advanced economies have become dominated by electronic money which has contributed to economic efficiency but has also converted information societies into debt societies. He believes that this has made the division between the modes of production between the developed and developing worlds more pronounced.

The emergence of such a new "information standard" or of electronic money has had profound effects on the international political economy to the point of challenging traditional notions of sovereignty. Pennings warns that new political and economic mechanisms for control will emerge to prevent technology-led democratization like the Internet.

In Chapter VII, David Lassner examines the development of international standardization and the impact this process will have on developing nations. He first provides a broad background survey of telecommunications standardization by analyzing the utility of standards and the dangers of inefficient standardization. His chapter demonstrates that standardization benefits users through interoperability, interconnectivity, decreased costs, portability, stability, and vendor independence. Vendors benefit through expanded markets, simpler network design, and reduced R&D costs. However, the international fora must act deliberately and carefully in the rush to standardize. As the backbone for advanced economic activity, the information superhighway must be interconnected. Standards must be for the long term because the costs of changing them are very high. The battle for competing modes of standardization for every newly developed technology can be intense and produces both winners and losers. As a result, Lassner examines two critical questions: who determines telecommunications standards and how are these standards developed? In his analysis, Lassner highlights the powerful emergence of regional standards organizations (RSOs) in recent years. Acting as intermediaries between national PTOs and the two leading international standards organizations, namely, the International Telecommunications Union (ITU) and the International Standards Organization (ISO), the regional organizations provide standards in those areas where international standards are not there or where only vague or optional guidelines are provided.

Finally Lassner examines the impact of these developments on the less developed countries which lack the time, money, and technically-qualified personnel to fully participate in the evolution of standards. Their interest has not always been represented in decisionmaking processes due to which long-term problems arise. The information superhighway will provide the backbone for the growth of economies in the future which implies that interconnectivity to the global telecommunications network is critical to economic growth. As Lassner argues, RSOs can play a crucial role in mitigating some of the disadvantages faced by the developing countries by giving them a greater voice in decisionmaking. However, it must be noted that unrestrained regionalism can choke off the efforts of global standardization and can do more harm than good.

Standards are equally important to the growth of the Internet and a discussion of its enabling power is the content of Chapter VIII. Starting with the evolution of the technology from a loose organization of computer networks used by scientists and academics into a global system with the introduction of multimedia, Jussawalla tackles the role of this global network in the construction of the GII. The Internet became not just a way to send e-mail, but a place which users visited, exchanged ideas, and gained the interactivity that broadcast media were trying to achieve: it created "cyberspace." With the arrival of the user-friendly multimedia side of the Net, the explosive growth of users began to enter the digital free market.

While it is not the advertised version of the information superhighway promoted by cable and telephone companies, it has the same basic ingredients of open flows of information and interactivity. Jussawalla argues that the Internet is already operating as the information superhighway because it is not centralized, it is more democratic, and it is overtaking on-line services with user-friendly home pages created by people who have some unique information to display. The Internet revolution allows everyone to be a publisher, creating billboards and participating in news groups. It has created thousands of cyberpreneurs profiting from home shopping, advertising, and entertainment. Accordingly its advantages outweigh its risks. It is a bottom-up information infrastructure as described by Rutkowski (1995). According to Jussawalla, the objectives of the Internet are its greatest advantages. It is designed to provide a seamless, affordable, information network that is not owned by any single organization, and it grows, like Topsy, without anyone knowing it. Its major asset is its World Wide Web, which has grown into a global organization that seeks to promote accessibility to as many users as possible.

Jussawalla cautions about its risks which users are already conversant with, such as safeguarding privacy of e-mail messages, credit card information, sensitive data, and unauthorized entry. Building "firewalls" and anti-spoofing software are creating a new market for software programmers. Managing the information overload is also creating problems of congestion on the Net. Probably a greater difficulty is the division between those who can afford to use the Internet and those who cannot, and this schism becomes more acute in a system that has become global in its outreach. The Net interprets censorship as unacceptable and devises ways to skirt around it. Its users are confident that in this electronic world, the system cannot be policed. The other side of the coin is that there are many difficulties in applying copyright laws to its users. The Telecommunications Reform Bill, which was passed in the U.S. Senate on June 15, 1995, tries to apply decency in communications rules to the Internet content, which is being hotly debated on the Net. Cyberspace may not respect the provisions of the law even though they are pretty stringent.

Examining the cost and price aspects of the network, Jussawalla deals with the difficulties of congestion on the conduits and the economic and social costs of waiting. Broad-band technology would reduce these difficulties. Internet's commercial potential has attracted the large telephone companies to roll out services with more bandwidth, security, and reliability with prospects of cost reduction and greater accessibility. But these plans have not yet come to fruition.

A novel approach to bridging the future is taken by John Locke in his chapter on telemediation in the fast-moving world of petabits. In Chapter IX, Locke believes that the future of telecommunications lies in its capability for multimedia information processing. He substantiates his hypothesis by tracing the roots of this transformation to the dynamism of information technology that makes delivery of data, voice, and video in larger quantities at faster speeds, i.e., transmission in quadrillion bits (petabits). Such a proliferation has resulted in a global marketplace which, according to Locke, is based on open systems integration. More and more, general purpose systems are replacing the older solution specific networks. In order to take advantage of this new "landscape," Locke argues that the large telcos, cable companies, and television moguls are merging in an effort to capture as large a share of the U.S. and global markets as possible. The various actors in this scenario are described as broadcasters and telecommunications providers participating in the new media revolution, but the challenge to them all is the lack of bandwidth. High-bandwidth, new media services will be required in the competitive infrastructure of the future. Into this scenario enter the data communications providers not just to support the operations of the "virtual corporation" but the requirements of the individual users.

While elaborating on practically every new technology available today, Locke keeps insisting that the successful model would be that of conversational, bidirectional interactivity. In conclusion, he reiterates his belief that the role of the consumer should not be bypassed by the overwhelming power of the conglomerates who supply the services, but should be a joint and cooperative structure in which the producers, carriers, and end users come together in the telemediation scenario of the future.

If Locke's vision is to become a reality for the global information infrastructure, what will be the role of satellites in contributing to the success of the future networks? This is elaborated on in Chapter X by Jussawalla. In this chapter, Jussawalla emphasizes the bold and innovative proposals from the commercial satellite-based versions of the GII and discusses the extent to which these have focused the attention of government and industry on the technical and social implications of a space-based superhighway. As traffic becomes more digitized, the satellites route it through an on-board switch. This has led to an explosion of satellite systems over the Pacific and Indian Ocean regions.

In the course of this chapter, Jussawalla examines the critical issues for privatizing Intelsat and Inmarsat in light of the market demand for satellites growing to $20 billion over the next decade. But for a global infobahn, mobile satellite services are the most effective way of getting people from different nations and cultures to participate in an interactive mode without sacrificing their national and cultural interests. Those countries that have vast expanses of rural dwellers without any means to communicate would find systems such as Iridium, Teledesic, Globalstar, and Inmarsat-P to be of help in linking them through the wireless revolution.

Jussawalla delineates the problems of orbital access as the race heats up for newer systems owned by country governments or private companies for domestic coverage of voice, data, and television and of regional systems like Asiasat, Palapa, and Apstar. Considering the numbers of GEOs registered for future launches, Jussawalla estimates that demand for cellular mobile systems and direct-to-home satellites will keep expanding. This will lead to a greater interdependence between satellites and cable network operators. This is likely because of the demand for multimedia services which will have a significant impact on emerging economies.

In the final chapter of the book, Mark Hukill raises critical questions about the policy focus for network economic activity. Like other authors in this volume, Hukill cautions about the widening gap in technology attainments between the developed and developing nations and then analyzes the kind of comprehensive policies that must address the different forms of economic activity that will ride on the information superhighway. In fact, Hukill's chapter rounds out all the technological issues dealt with separately in this volume and provides a framework for assessing the kinds of policies that will be required in the future to make access to the superhighway more equitable than it currently is. In the course of this discussion, he specifically deals with three separate networks—private data networks, PSTN, and Internet—and their potential to deliver interactivity. This leads to the interdependence of economic activity and technology development which, in turn, determine the kinds of policies that will be formulated.

An interesting model is built by Hukill to demonstrate the bimodal relationship between business-related economic activities and network activities. He suggests that pricing policies of information services need to be reexamined in the context of network expansion and globalization of their impact. A series of questions are raised for policymakers in the various areas of competition, universal access, anti-trust, trade, tariffs, and copyright which are all interrelated but can stand alone as well.

This entire volume of eleven chapters has dealt with the growth and impact of new technologies on the marketplace of ideas which will be affecting the corporation, the workplace, the home, schools, and the health of societies and individuals in the coming century. While information will flow effusively over faster conduits, it is how this information is absorbed that will determine the prosperity of future generations. Economic progress will occur only when the powerful tools of information technology allow people to share their knowledge. It is not enough to transfer or to duplicate files of data, but societies and people must be able to create meaningful patterns from them and transfer real knowledge. According to economist Paul Romer, technology is now an "endogenous" part of economic activity and not an "exogenous" one as described in traditional economic theory (Robinson 1995). Many decades ago, Hayek (1945) in his classic essay, *The Use of Knowledge in Society,* stated that knowledge does not exist in concentrated or integrated form, but as dispersed bits of incomplete and contradictory information from which knowledge has to be gleaned.

Secretary General of the ITU, Pekka Tarjanne (1995), wisely cautions that in the global information society of the twenty-first century, there should be no room for centralized control and that interactivity is the most important attribute of democratic societies. He further elaborates that "from the perspective of today's realities, the development of the communication networks of the future cannot—and should not—be separated from overall human development" (p. 28).

REFERENCES

Business Week, 1955, "The Future of Money", June 12, pp. 66–70.

Hayek, F. A., von, 1945, "The Use of Knowledge in Society", *American Economic Review*, vol. 35, no. 4 (September), pp. 519–30.

Marks, Leonard, 1995, "Detours Along the Information Highway", *I-Ways*, (January/February), p. 38.

Mitchell, Russell, 1995, "The Key to Safe Business on the Net", *Business Week*, February 27, p. 86.

Robinson, Peter, 1995, "Interview with Paul Romer in ASAP", *Forbes* (June), p. 67.

Rutkowski, Anthony, 1995, "The Five Faces of the Internet", IIC Communications Topics Series, International Institute of Communications, London, February.

Tarjanne, Pekka, 1995, "The GII Moving Towards Implementation", *Telecommunications* (May), p. 28.

FROM THE NETWORK OF NETWORKS TO THE SYSTEM OF SYSTEMS

Selected Countries
L96
L98

ELI NOAM

Telecommunications is moving inexorably towards competition, deregulation, and broadbanding, and as it has done so, much has been discussed; but the most fundamental questions of policy are rarely asked. And they are, after competition, what? After deregulation, what? And after broadbanding, what? Most observers focus on present bottlenecks— psychological, regulatory, and financial—and not on what is being bottled up. In the United States, the day is not far off, historically speaking, when entry will be wide opened—when fiber is widespread in all stages of most networks, when radio-based carriers fill the still-substantial wide-spots map of telecommunications ubiquity, when foreign carriers would operate in America and vice-versa. In such an environment, what market structure can we expect? What problems do we expect to face and what regulatory environment must we erect? We all know that the monopoly system has passed; that is, the single carrier has given way to a network of networks, to an environment in which multiple carriers interconnect into each other foremost in the long-distance field, domestic and international, but also increasingly in the local field and in mobile. And these networks provide the elements for the matrix of a network of networks that envelopes us electronically. But what about the ability of the actual user, the end user, to handle the newly established, organizational environment that is totally different from the simple single monopoly?

There is a great need for an integration of the various parts of the network in the future. There are several ways to integrate, and the first way to integrate is by end users. And those can be either integration through do-it-yourself users, which is basically today's system for American residential-users, in which the users arrange for their own long-distance company and their own terminal equipment. Large users also put together their own networks and their own leased-lines operating systems. My own university, Columbia, for example, does this, employing forty-five people in the process of its own network integration. But self-integration gets complicated very rapidly. Even large users such as Citicorp choose not to integrate themselves anymore but to outsource, so why should a small residential user not be able to do so? Secondly, terminal-based integration assists a user's terminal equipment and incorporates such built-in intelligence which can make the right choices among carriers and service providers on a real-time basis. The PBXs of large corporations have had, for example, a least-cost routing option. And this concept has been extended to the residential market in Japan by one of the long-distance competitors, DDI, which has persuaded millions of the Japanese to buy special terminals that can automatically pick the cheapest carrier for any given call. The underlying database is supplied by DDI itself, which not surprisingly is usually the cheapest carrier; thus DDI wins, though it could lose much of its market overnight if another entity were to underprice them. On the whole, customer-premises integration, even though done through intelligent devices, requires significant transaction costs and people have difficulty handling it. This gets us to the second option of carrier-based integration which comes in several varieties. One is the expansion of facilities providers—telephone companies, for example—in end-to-end carriers that offer everything, and this can be by horizontal expansion or by vertical expansion into new services. It is not realistic to imagine today that any company can be big enough and varied enough to offer all four kinds of services and facilities. And to do it well, locally, domestically, and internationally across services is difficult to accomplish. However, joint ventures among carriers is, of course, possible and, in fact, is happening. It is a likely scenario and one which is emerging.

The third option is integration by systems integrators. Perhaps the most promising and innovative scenario for the integration of the bits and pieces of networks is systems integration. A new class of systems integrators is emerging. Their role is to provide the end-user, typically corporate, governmental, or family groups, with access to a variety of services in a one-stop fashion. These specialized integrators also known as outsources or managed data service providers assemble packages of services, tariffs, and hardware customizing these packages to the special needs of their customers. Systems integrators have always existed in other fields, for example, in general contracting, construction projects, or in travel packages, and for computer service firms which were probably the direct forerunners of telecommunications systems integrators. The characteristic of pure systems integration is that they do not own or operate the various subproduction activities but draw the select optimal elements in terms of price and performance, package them together, manage the bundles, and offer them to the customer on a one-stop basis. This relieves customers from the responsibility of integration for which expertise is required and yet they are not obliged to recover major investments as carriers are. Now who will be systems integrators? A whole bunch of people; they can be local exchange carriers, cable television companies, long-distance carriers, international carriers, or telecommunications resellers, in particular, computer systems providers value-added service providers. Similarly, office automation firms, land providers, high-tech firms, defense contractors seeking diversification, and corporate networks with excess capacity can be systems integrators. Today, systems integrators exist for large customers and they have also begun to be active in establishing group networks, establishing internetwork telecommunications. But when it comes to small users, Mom and Pop need not apply, though things may be quite different tomorrow. Additional staff would be needed for systems integrators to merge and put together individualized network for personal users, establishing personal networks.

Before dismissing the notion of personal networks as extravagant, remember that only a dozen years ago nobody expected personal computers on everybody's lap either. What does a personal network mean? It means an individually tailored network arrangement that fits an individual's telecommunications needs. It doesn't necessarily mean separate physical systems, but mostly virtual arrangements with various bandwidth on demand for various service options including details of access numbers, data processing, bulletin boards, transaction programs, videotex, data sources, as well as traditional telecommunications. Now these packages will be provided by systems integrators as systems and these systems, in turn, are going to interconnect and interaccess with each other, sprawling across carrier service providers and national frontiers. In so doing, the telecommunications environment evolves from a network of networks, which is a physical arrangement in which carriers integrate, to a system of systems in which systems integrators link up with each other.

Now, where does that new environment leave government regulations? In the recent past, policy debates centered on the opening of telecommunications including television and cable television. Is competition sustainable? Is it advisable? Who gains? Who loses? Regulations have been essential to the old system, partly to protect against a monopoly and partly to protect the monopoly itself. Now, can one expect this system of systems to be totally self-regulating with no role for government? Does liberalization imply libertarianism? The option of an invisible hand mechanism—the ideal that out of numerous decentralized sub-optimizing actions there would emerge overall beneficial equilibrium—and its importance goes way beyond economics. Can electronic communications function in such a fashion, optimally arranging themselves in the absence of an overall plan or direction? The mere notion is almost incomprehensible to telecommunications traditionalists. Why do we have regulations in telecommunications? Not for the self-interest of bureaucracy per se, but ultimately there are some underlying policy goals. They include universal coverage, affordable rates, free flow of information, restriction of market powers, restriction of monopoly, pricing, effectiveness, support of high-technology interconnectivity in society, interconnectivity of equipment, privacy, security, etc. And to assure those goals, most of which are of the apple pie variety, regulators instituted a variety of policies such as rate subsidies, universal service obligation, common carriage, interconnection rules, access charges, quality standards, limited liabilities, and others.

In a system of systems integrators, however, what forms of such regulations are still necessary and what new ones if any may be required? Regulation by government existed partly to right the imbalance of power between huge monopoly suppliers, on the one hand, and small atomized and technically ignorant users of services, on the other hand. In a system of systems, integrators act as the user's representative or agent vis-à-vis the carriers. They can protect the users against a carrier's underperformance in quality or underperformance in privacy or too high prices. This assumes that users have a choice amongst systems integrators. In other words, if the systems integration market itself is competitive and the systems integrators have a choice among noncolluding suppliers of underlying services, they can protect users. Of the policy goals which are affected by the system of integration, price, quality, market powers, security, and even privacy are handled by systems integrators. On the other hand, policy goals that are left unresolved and potentially in need of regulatory support are universal service, affordable rates, free flow of information, and interconnection. But in the future, this system cannot be maintained; the new flow will have to become externalized rather than internalized and probably will have to be replaced by different systems such as charges, a form of value-added taxes, or some other form of generating revenue which would go into a universal service fund that could be used to support certain network functions providers, categories of users, or carriers. Thus, government is not likely to disappear from this area.

The remaining problem is the free flow of information. This was, in the past, handled by common carriage which is an arrangement of free access, nondiscriminatory access, nondiscriminatory neutral use. But in a competitive system, that arrangement will change. Systems integrators are not common carriers; they can include or exclude whom they want to. Columbia University, for example, can restrict certain forms of communications on its network. If somebody resells communications packages to some customers, they can impose conditions on it. They can sell those packages to some customers and not to others. It is doubtful if this is going to change. I don't think that systems integrators will become common carriers and therefore, the interconnectivity of systems integrators will also not be based on a common carriage system. The underlying carriers could remain common carriers, although that's a question too. The system as a whole is not going to be one of a common carriage system, which means that the carriage of certain forms of communications, particularly of the unpopular variety, is going to be jeopardized, and that means that certain nondiscrimination rules might still have to be required by governmental regulations.

Interconnection itself must interoperate in terms of technical standards, protocols, and boundaries. Yet interconnectivity does not happen by itself; this is the lesson of the American experience. An open network architecture that has a comparably efficient interconnection collocation is part of that evolution. Issues of regulations that the system of systems raises are integrator power and their policies. Is it possible that only a few integrators might survive if there are strong economies of scale and scope leading to an oligopoly or even a monopoly in systems integration? This is unlikely. It is unlikely that systems integration is a natural monopoly. The economy of scale that could lead to market power might be an ability of large systems integrators to get advantageous rates from carriers. But, generally speaking, it is hard to imagine that economies of scale are similar for each of the different services or for the layers and the hierarchy, and therefore, it may not happen.

Now, it is possible that systems integrators could have excessive power over the customers once they sign up with them. Again this is not likely; customers could move away and competition would work here too. So I conclude it unlikely that systems integrator power will emerge. However, a more real possibility is that the underlying carriers will have excessive powers if they also function as systems integrators, an issue that is not problematic, but where they also maintain bottleneck power over some segments of their network, in particular, in the case of local exchange companies. Carriers that function as systems integrators could favor their own segments of service or equipment.

In a competitive environment, it is more likely that independent integrators will have a competitive advantage over established companies who promote their own services over lower-priced independent suppliers. To be truly competitive as a systems integrator, a traditional carrier systems integration operation must be willing to compete against its own carriers and, in effect, become independent. And while that is conceivable, it might require significant rethinking. In addition, it will be easier for new firms to enter if they do not have to build end-to-end networks and can resell a bulk of the network of others if they can have others market their own production. A systems integrator should enhance competitive entry for carriers. This is not to say that traditionally carriers do not have some advantage for they do, including coordination of planning, advance information, established goodwill, and reduced transaction costs for operations under one corporate rule. Carriers with strength in these areas might therefore establish themselves as competitors in systems integration. And then, of course, it is possible that a carrier could provide preferential services to its own systems integrator. Though in a competitive environment, it is not rational economically to limit one's sales to one's own outlet. In a way, market power exists in the carriers service segment. Regulation should assure nondiscriminatory services. So where monopoly power persists in any transmission segment, competitiveness would have to be assured by imposition of nondiscriminatory access to these segments but with standby safeguards available. There should be no problem of having carriers of dominant systems integration.

A new problem that arises is international asymmetry. The system of systems works as long as it is competitive in each of its stages or as long as regulation establishes nondiscrimination. However, in an international setting, neither of these conditions is likely to be met. Most countries lag behind the United States and Japan in the evolution of networks. They are only now slowly entering the stage of multiple facilities–based carriers. The traditional monopoly carrier is almost always firmly entrenched, operating in all stages of communications. In consequence, systems integrators cannot truly compete against the semi-official telephone organizations in systems integration except in each domestic market. This might be considered to be an internal problem which has a global anti-competitive impact. This is the case because some of these telephone organizations are aggressively pursuing international systems integration themselves, while at the same time holding gatekeeping powers over entry into their own home markets. If a PTO of an important European country could restrict the effectiveness of an American or Japanese systems irregulator to offer global services, while at the same time entering or liberalizing environments in the United States and Japan themselves, it could also operate to benefit the interest of allied equipment manufacturers. Of course, other countries' PTOs can play the same game and as a result a new trend of international carrier collaboration has emerged in which major PTOs enter into joint ventures of systems integration. Examples are Syncordia, Infonet, Unicom, and Unisource to mention just the best known. Potentially at least these alliances of dominant national carriers could create international cartels in various countries to compete with other systems integrated. Whether in their home countries or in the international arena, this has an anti-competitive potential of whipsawing in which one side of liberalization encroaches frontiers and commits the remaining monopolies to fully appropriate previously-shared monopoly profits. This is similar to

what has happened in long distance or what is feared will happen in the
long-distance field. To prevent this, it is essential to press internationally, for
nondiscriminatory access, lease and interconnection arrangements that are
neutral to the nature of the nationality of the systems integrators. And the
United States and Japan, being the largest and most interesting markets for
systems integrators, should exercise leadership in pressing for such
reciprocity.

In the 1980s, telecommunications policy was centered on open entry.
This was correct then and now, even if it was often unpopular. But in the
1990s, there must be a different emphasis. In this decade, issues of
integration of the various network parts will be at the forefront. This is, so to
speak, the post-deregulatory agenda. And for that reason, when I was in the
New York Public Service Commission, I concentrated on integration issues
such as common carriage, multicarrier ISDN, ONA, collocation, quality, and
privacy because I believed that these issues, as they were being experimented
with on the state level, will and already have come to the national forefront. It
is essential to reconcile the centrifugal pressures that have emerged in the
network of networks with a need to interoperate and intercommunicate. And
this is the main challenge for policymakers for the next decade. This means
we must provide the competitive system with tools of interoperation, where
they are not self-generated by market forces, and to deal with the institution
of integration. Systems integrators will emerge, as I argue, as the central

elements of future telecommunications environment. In the United States, the past decade has been preoccupied with market liberalization. This will continue but it will also be inevitable to move beyond this agenda and to assure the function of a system of systems.

THE IMPACT OF REGIONALIZATION ON THE FUTURE OF EMERGING MARKETS IN INFORMATION TECHNOLOGY AND TRADE

DON LAMBERTON

A pattern of thought has been fueled by spectacular technological developments but also stems from considerations of self-interest. It was that great Cambridge, U.K. economist Joan Robinson who said it should occasion no surprise that free trade was advocated most ardently by the country that stood to gain the most. My view is that her thesis can be extended to telecommunications interests when they proclaim the seamless world, albeit one in which, according to the ITU, half the world's population does not live within two hours' walking distance of the nearest telephone!

This application of the idea of the global village promotes some of the self-interests involved, lends an aura of excitement over new devices, and attracts many who run with the hounds. So I start from the premise that nothing can safely be considered sacred. I will suggest that the reality is regionalization and that the "globaloney" school of thought errs on such crucial matters as the sources of comparative advantage in a dynamic world, the benefits of integration, and the nature of international competition. Their analysis, which reflects both self-interest and the persuasiveness of the telecommunications technocrats, treats the global economy as an attainable state, whereas we might expect it to recede before us like the economists' stationary state and other utopias. The integration process that unified the United States and Australia, for example, does not provide a model for the world economy.

I have previously described 1945 as a good vintage for information society buffs and especially for IT ones, even if few then had the vision to take full advantage of the rich mix of perceptions that were being offered. For example:

1. Arthur Clarke's paper, "Extra-Terrestrial Relays Can Rocket Stations Give World-wide Coverage?", which was published in *Wireless World*, showed that satellite communication was possible. Implementation and diffusion of this technology probably did more than anything else to foster the global myth.

2. Hayek's "The Use of Knowledge in Society" had appeared a month earlier. It was a significant milestone on the road to the development of information economics, purporting to show that the market system could handle the use of knowledge in society.

3. In mid-1945 Vannevar Bush sprung the notion of information overload upon us in his essay, "As We May Think." He pointed to the growing mountain of research, the proliferation of specializations and the importance of inventing new ways, e.g., his memex, to ensure that scientific information was used.

So half a century ago, the basis had been laid for much current

thinking about a partnership of technological advance and the market

system.

1.   WHAT IS GLOBALIZATION?

Against this background, let me pose the question: What is

globalization? One approach emphasizes capital flow. True globalization

would mean that capital would flow in a roughly uniform way into all

developing regions, being attracted by comparative advantage: "there should

be no pockets of stagnant growth where disciplined work forces and/or

abundant natural resources are under-utilized" (Johnston 1991, p. 1). Quite

clearly some areas receive more economic attention than others, so world

markets are not global according to this test: "closer examination of the flow

of capital, trade relationships, and income levels reveals strong regional

patterns of development. Thus regionalization frequently appears to be a

better description of world market evolution than globalization" (Johnston

1991, p. xi). Projection of these trends can be interpreted as a shifting of the

balance of economic power away from the United States and toward Europe

and Asia, even an Asian future in Europe as presented by Gunter Grass in

his latest novel, *The Call of the Toad*, with the bicycle rickshaw, produced in

European plants, solving the modern city's most pressing problem.

Johnston examines developments in Japan, the rest of Asia, Europe, the Soviet Union and Eastern Europe, and sub-Saharan Africa before turning to a rethinking of U.S. business prospects. She succeeds admirably in substantiating the unevenness of capital flows, but there is much more than the flow of investment funds involved. Her test of globalization applies to areas similarly endowed. Can the complex mix of money capital, disciplined/trained and not-so-disciplined/trained workforces, known and imagined natural resources, and cultural traditions, along with varying degrees of political stability and social cohesiveness and different patterns of external ties, be disposed of so easily? Together these constitute the social capabilities of the nation; they are its organizational capital. Few areas will, therefore, be similarly endowed. These considerations strengthen Johnston's argument: i.e., that investors are looking for profit and would be unwise to ignore the lack of such capabilities.

Information, communication, and coordination aspects dominate in the integration process. If we treat the information resource as shorthand for the information stocks and flows, the capability of handling that resource, and the institutional channels that have been created for that handling, then there is reason to think, first, that an existing distribution of information resources may confer a comparative advantage on a regional bloc, and, second, that regional bloc activity may achieve further advantage through specialization in information activities.

Some of the resources needed to achieve such an advantage are not readily amenable to GATT-type leveling of the playing field. Consider the case of culture. If by culture we mean the living historical product of group problem solving, it is a public and private sector form of organizational capital that will not be captured in any GATT talks on investment and yet can confer comparative advantage. Perhaps the sharing of such organizational capital can be a source of sufficient benefits to make regional integration more attractive than the shared benefits on a level playing field peopled by players of very different sizes and capabilities.

Of course, attempts to carry leveling to the extreme may undermine the world trading system and lead to a loss of benefits. Bhagwati (1991), for example, has argued recently that the entire system is at risk if the United States extends accusations of 'unfair trade' practices to areas as diverse as retail distribution systems, infrastructure spending, saving rates, and workers' rights. I recall the late Fred Hirsch's remarks about the question-begging character of the concept of equal opportunity. For him, "the central ambiguity" was which starting handicaps were to be removed." At the limit, the criterion of an equal start is an equal finish" (Hirsch 1977, p. 5).

When Johnston turns to possible solutions of the U.S. problem, she calls for a more competitive stance by U.S. manufacturing—with the United Auto Workers possibly starting a new auto company to produce a small car.

She believes special attention will have to be given to Latin America to stem migration before its impact becomes intolerable; it is here that we come to a real weakness in the global view. The theoretical model underlying the idea that free market forces should guide development calls for free movement of resources, and yet there seems no prospect that the people wanting to climb the economic gradient by migrating to richer countries will be allowed to do so.

Let me turn to other ways of thinking about a global system. The trading possibilities that are fundamental to the functioning of an economy arise from individual differences. These asymmetries, especially those of information and organization, must be viewed in the context of the global economy. The word "global" means pertaining to or embracing the totality of a group of items, categories, or the like. In relation to economic activity, the term "global" would seem to imply, at a minimum, the involvement of all economies. But that involvement could well be limited, both in terms of the volume and variety of trade. At the other extreme—and this would seem to fit well with the popular global village thinking—there would be complete economic integration. Can there be anything that might be described as complete integration while individual differences of information and organization remain? Perhaps we should see the global economy as a set of individual economies showing diversity with respect to both position within this integration spectrum and overall size.

1. In one of the few theoretical papers that explore the workings of a global economy, Holland (1988) notes the following features:The overall direction of the economy is determined by the interaction of many dispersed units acting in parallel. The action of any given unit depends upon the state and actions of a limited number of other units.

2. There are rarely any global controls on interactions. Controls are provided by mechanisms of competition and coordination between units, mediated by standard operating procedures, and assigned roles and shifting associations.

3. The economy has many levels of organization and interaction. Units at any given level typically serve as "building blocks" for constructing units at the next higher level. The overall organization is more than hierarchical, with all sorts of tangling interactions (associations and channels of communication) across levels.

4. The building blocks are recombined and revised continually as the system accumulates experience; that is, the system adapts.

5. The arena in which the economy operates is typified by many niches that can be exploited by particular adaptations; there is no universal supercompetitor that can fill all niches.

6. Niches are continually created by new technologies and the very act of filling a niche provides new niches.

7. Because the niches are various and new niches are continually created, the economy operates far from an optimum. Said in another way, improvements are always possible and, indeed, occur regularly.

8. In some ways, e.g., the numbers criterion, Holland's model seems to be the competitive model writ large, but it can be adapted to accommodate real world aspects of structure and competition. While he has many dispersed units acting in parallel, what an individual unit does depends on what a limited number of other units do. If the units are nations, we may have regionalization; if they are business entities, the limited number would imply oligopoly and we should look at conditions of entry. (I'll come back to this matter of competition and industry structure.)

Several other features of such a model might be noted in this context. First, there is an important role for accumulated experience. Experience and learning are the basis of improved standard operating procedures that, in turn, permit more sophisticated interactions between units. If culture is "a living historical product of group problem solving" (Van Maanen and Barley 1985, p. 33), it readily finds a place in this analytical framework. So too would a theory of learning.

A second feature relates to limits on adaptability. Reliance upon accumulated experience can, because of the economic characteristics of information and the organizational arrangements for information-handling, lead to failure to satisfy and a failure to search for better outcomes. Culture, as defined above, ceases to be part of social capital and becomes a burden. History matters and the process leads towards organizational obsolescence (Arrow 1974, p. 49).

A third feature concerns technology. There is presumably a continuing supply of new technology to ensure the perpetual novelty. As this does not come from outer space, the model needs to provide for its creation and to insure that there are conditions permitting technology transfer. Both the creation and transfer phases are complex and costly, so they open up opportunities for monopolization.

Assigning an organizational, information-using character to the economy and its operational units may have brought greater complexity to our analysis, but it may have been a price worth paying. An information-theoretic viewpoint allows a role for learning, the design of organization, sequential behavior, time lags, and consequential diversity. In contrast, mainstream economics does not have a perspective on technology and the future of the information and communication industries. Insofar as it has failed to recognize limitations upon our capabilities to gather and process information, to communicate, decide, and to agree amongst ourselves, mainstream economics has in effect denied the possibility of any special role for these industries. Of course, economics has gone even further than such a denial: its most polished theoretical models are still based upon assumptions about perfection, including perfect knowledge. From those models, one might be led to predict a zero resource allocation to information activities. Clearly, someone is in error because those activities taken together—i.e., information processing, interpretation, and intermediation—are now the dominant resource-using activity. Thus, mainstream economics working with given tastes and given technologies does not seem to hold much promise.

There may be objections raised to my charges of neglect: imperfect information is now said to be a standard consideration of economic theorists. In a superficial way this is true. However, what has not been achieved and what is clearly needed is a modeling of the economy, or of its parts, "in which information is continuously being collected and processed and in which decisions, based on that information, are continuously being made" (Stiglitz 1985, p. 23). To do this, I contend we must look to the informational and organizational features, which, to the extent they are thought about at all in the mainstream tradition, are bundled together with roads, dams, bridges, power lines, airports, telecommunications systems, laboratories, and institutions and given the label "infrastructure." Infrastructure may be a guiding concept in the vision of the new America, but it is "largely a symbol, a metaphor capable of generating some pretty fancy daydreams" (Muschamp 1993, p. 32).

Infrastructure has to include organizational capital. There is currently some interest in the concept of intangible investment and a sense that by concentrating on equipment and buildings, we may have missed some important pieces in the productivity puzzle. A workshop at the OECD in December 1992 and a session scheduled at the American Economic Association meeting in 1993 focused on this topic.

I am suggesting that both the narrow, capital flows notion of globalization and the more elaborate Holland model lead to these considerations, considerations that have an important bearing upon any attempt to design organizations, whether they be institutions, cities, or regional blocs. The new technologies, especially the new information and communication technologies, are contributing to change, eliminating choices, and forcing new choices on individuals, business firms, and nations. The nature of these processes is not well understood. A major danger is that problems are seen as merely technical with consequential neglect of the human, social, and economic aspects. Large parts of the literature are distinctly unhelpful. The utopian envisages a better world but says little or nothing about how to get there, the rhetorical serves to keep technology-based hopes alive under increasing economic pressures, and a sales effort emanates from firms with ICTs for sale or from those, some scientists included, whose livelihood seems dependent upon adoption of these technologies.

I go on to say that these considerations are the very essence of economic dynamics and the information economy, and that they link to the big questions about comparative advantage.

2.   COMPARATIVE ADVANTAGE

Much of the history of the international economy over the last half century is the application of the tenets of economic liberalism: free play of market forces, a limited role for government, free exchange and interest rates, openness to imports of goods, services, and capital of other nations, and integration into the world economy.

The key proposition underlying this doctrine is that the appropriate international division of labor and the realization of a country's comparative advantage are best achieved by free market forces. However, as Penrose (1992) asserts, "neither appeals to history nor experience demonstrate its truth in a dynamic world. Where did the government not play, in collaboration with local business and other elites, a very powerful role?" (pp. 238–39). Not one, says Penrose, got it right by relying on the free play of market forces, abandoning subsidies, and being freely open to imports and foreign investment. Leafing through the pages of history, she contends that what was true of the British economy and North America holds for the non-European countries that successfully industrialized today.

Where then is the experience that justifies what she calls "the repeated insistence" of the IMF and the World Bank that developing countries should practice this economic liberalism? One of her most telling points relates to the Bank's assertion that while industrial countries have restricted trade to promote industry, their objectives could have been attained in other ways. The Bank asks that the standard of demonstrability be applied to justify departures from the liberal style policy, but it does not attempt to apply that standard to its own theorizing.

3.   ECONOMIC INTEGRATION

The role of integration as an aspect of economic liberalism for developing countries requires careful examination. Imports and technology may flow in with investment without achieving lasting, effective technology transfer; local resources may be diverted from national priorities; and profits may go abroad. The contrast between the ideal and the reality arises because the long chain of assumptions on which the liberalism case rests breaks at several points:

3.1. Industrial Structure

Integration would be a paying proposition if the structure of the major international industries were an open one of the textbook variety, with easy entry. The reality is that there are barriers to entry, concentration, and linkages that create a strong industrial fabric, interweaving business and political power. In this area more than in any other, Herbert Simon's claim that what happens in organizations can be even more important than what happens in markets is applicable (Simon 1991). And I should add that this combination of features is not just the latest in organizational fads: consider, for example, the history of the grain trade.

What are these features? International business is now characterized by:

> rivalry in highly personalized markets managed by advertising and other techniques; it specializes in new products, new technology, new 'image-making' for consumers among a few very large, often also fairly evenly matched, multi-corporate conglomerate enterprises, global [worldwide] in scope and with a great deal of political as well as economic power...[These enterprises scan] the world for commercial opportunities like satellite receivers. None is a single corporate company, or enterprise, but is rather a cluster, or even several clusters, of many kinds of enterprises, corporations or companies, often with intricate tiers and lines of ownership, responsibility or authority...[It is a system that] combines principles of co-operation and competition realizing the importance of both, within an overall framework (Penrose 1992, p. 241).

3.2. Technology

Not only does this business system derive strength from the gathering and pooling of market information, but the alliances extend to things technological. The firms therefore have a competitive edge in the use of business intelligence to guide the creation of technology, the transfer of technology for their own purposes, and the utilization of that technology. This edge is enhanced because all of these activities become routine operations, while for the developing country that is a new entrant, these are all initial experiences.

3.3. Infrastructure

The role of accumulated experience and the stocks of information it can yield point to the importance of private-sector organizational capital, which in the information economy is a necessary complement to the information and information-handling facilities provided by the public sector. Despite the tenets of economic liberalism, governments have retained a significant role in information provision, education, research, and coordination, and the maintenance of power both domestically and internationally. As a consequence, one of the major alliances is between government and industry. In a very real sense, the costs of competition internationally have been shifted to government because governments compete in the provision of infra-structure. This would appear to be the case of telecommunications. Not only do governments themselves need these services, but industries are obliged to use them to conform with international practice and their absence will be a deterrent to foreign investment.

> We should not be surprised at the finding that:

> Telecommunications is reinforcing the commanding role of major business concentrations around the world...New information technologies do have a fundamental impact on societies, and therefore on cities and regions, but their effects vary according to their interaction with the economic, social, political, and cultural processes that shape the production and use of the new technological medium (Castells 1989, pp. 1–2).

While developing countries may appear to gain some advantage by leapfrogging technology, they may miss the opportunity because of financial pressures, competing needs, and the speed of adjustment.

It is the possession of infrastructure and requisite social capabilities that divides nations into two groups: the "convergence club" and the rest of the world (Nelson and Wright 1992). The much talked about international-ization is very largely limited to the "club." The entry requirements are stiff and the dues must be paid over a long period. This is a slow process that involves the creation of "a well-educated workforce including competence at the top in the major sciences and technologies of the era, adequate firm management and organization, and financial institutions and governments capable of keeping fiscal and monetary houses in order" (Nelson and Wright 1992, p. 1949).

These capabilities have to be created and maintained as a package. This organizational capital is the real infrastructure, the real wealth of nations. (I have examined this concept in more detail in a chapter in a forthcoming book edited by Robert Babe.)

3.4. Labor

The market approach finds it difficult to reckon with the most conspicuous factor immobility, that of labor. Provisions for political asylum do not seem able to be stretched as easily as could, for example, trade restrictions of the CoCom variety. If trade flows are to be free, why not freedom for labor movement? Penrose adds to her criticism of the World Bank noting that on this important topic, the *1991 World Development Report* had but one paragraph stating that labor mobility could reduce the disparity of incomes and reduce population pressure.

There is a reference I should like to mention here because it contributes to the historical perspective. It was a study of population aspects of the Industrial Revolution. I can recall only the subtitle: "Shoveling out Paupers"! And the outcome to date in the German experience of integration has been described by the *Wall Street Journal* as being more similar to a den of foreboding rather than a land of opportunity, even though East Germany can now boast big-screen TVs, nine channels, and remote control.

56

3.5. Regionalization

I have attempted to challenge a widely prevalent view of trade and development strategy and I hope this can be brought to bear on the IT issues. Perhaps these regional patterns will prove to be incidents on the way to the global village, although I suspect not. Countries faced by the realities of the international economy but with strong development aspirations, reach out on the basis of their experience and their limited knowledge. There is no recipe book to be brought by visiting experts for the creation of the requisite capabilities. They are conscious of a variety of costs: the costs of thinking, of agreeing, of foresight. They must reckon with a variety of constraints: psychological, cultural, and organizational. The pattern of incentives and costs favors "production...becoming more local and regional; trade...becoming more continental; and capital and currency markets...becoming more international" (Bell 1990, p. 229). As noted in the *The Economist* (1990), "financial markets may indeed have turned global, but that does not mean that every financial intermediary must build a costly worldwide network. The business and the variables that determine competitive advantage are more complicated than that" (p. 16).

Nobel prizes create new fashions and Coasian economics with its emphasis on the costs of using the market is no exception. For countries that do not belong to the "convergence club," the strategy may well be to try to keep costs down by building their alliances through a mix of traditional ties, common culture, and the best deals offered by foreign investors. Language costs are important even if advances in machine translation are being made. Eighty percent understanding of another language is not good enough when dealing with the fine details of legal and scientific matters, although it will create more jobs for lawyers.

The implications for telecommunications and IT more generally would seem to be that they will continue to reinforce the important role of major business concentrations in the world today, to the probable advantage of the convergence club. Interconnectivity can cater to those demands that are worldwide—finance, stock markets, and the like—as other services have done in the past. Underlying cost conditions will sort out the problems of standards, with the aid of the ITU, and impose a check on those 'pretty fancy daydreams.'

Last year I asked an audience to name the global firms. The only one on which there was any measure of agreement was IBM. I suspect there would be even less of a consensus now.

REFERENCES

Arrow, K. J., 1974, *The Limits of Organization*, Norton, New York.

Bell, Daniel, 1990, in R. Swedberg, *Economics and Sociology Redefining Their Boundaries: Conversations with Economists and Sociologists*, Princeton University Press, Princeton, N. J.

Bhagwati, J., 1991, *The World Trading System at Risk*, Princeton University Press, Princeton, N. J.

Bush, Vannevar, 1945, "As We May Think", *Atlantic Monthly* (July), pp. 101–108 as reprinted in A. E. Cawkell (ed.), *Evolution of an Information Society*, ASLIB, London, 1987.

Butler, R. E., 1988, "Deregulation in the 1990s", *International Telecommunications Union*, March 8.

Castells, Manuel, 1989, *The Informational City: Information Technology, Economic Restructuring, and the Urban-Regional Process*, Blackwell, Oxford.

Clarke, Arthur C., 1945, "Extra-Terrestrial Relays Can Rocket Stations Give a World-wide Radio Coverage?", *Wireless World*, LI (October), pp. 305–308.

The Economist, 1990, "Computers, Telephones, Tunnels," December 15, p. 16.

Hayek, F. A. von, 1945, "The Use of Knowledge in Society", *American Economic Review*, vol. 35 (September), pp. 519–30.

Hirsch, Fred, 1977, *Social Limits to Growth*, Routledge and Kegan Paul, London.

Holland, J. H., 1988, "The Global Economy as an Adaptive Process", in P. W. Anderson, K. J. Arrow, and David Pines (eds.), *The Economy as an Evolving Complex System*, Addison-Wesley, Menlo Park, California, 1988, pp. 117–28.

Johnston, H. J., 1991, *Dispelling the Myths of Globalization: The Center for Regionalization*, Praeger, New York.

Lamberton, D. M., 1991a, South Pacific Economic Integration, paper presented at the North-East Asia Round Table II: Prospects for NorthEast Asian Economic Integration, Seoul, October.

———, 1991b, The Information and Communication Industries' Globalization and Regionalization: The Current Trends and Future Prospects, paper presented at the Third KISDI (Korea Information Society Development

Institute) International Conference: Globalization, Regionalization and Informatization, Seoul, November.

————, 1992, "Regional Economic Integration or the Global Village?", in *PTC' 92 Conference Proceedings*, PTC, Honolulu.

————, 1993, "The Technology-Human Factor Relationship", in *PTC' 93 Conference Proceedings*, PTC, Honolulu.

————, forthcoming, "The Information Economy Reconsidered", in Robert Babe (ed.), *Information and Communication in Economics*, Kluwer, Dordrecht.

Muschamp, H., 1993, "Thinking about Tomorrow and How to Build It", *New York Times*, January 10, Section 2, pp. 1 and 32.

Nelson, R. R., and G. Wright, 1992, "The Rise and Fall of American Technological Leadership: The Postwar Era in Historical Perspective", *Journal of Economic Literature*, vol. 30, no. 4 (December), pp. 1931–964.

Penrose, E. T., 1992, "Economic Liberalization: Openness and Integration but What Kind?", *Development Policy Review*, vol. 10, pp. 237–54.

Simon, H. A., 1991, "Organizations and Markets", *Journal of Economic Perspective*, vol. 5, no. 2 (Spring), pp. 27–35.

Stiglitz, J. E., 1985, "Information and Economic Analysis: A Perspective", *Economic Journal*, supplement to vol. 95.

Van Maanen, J., and S. R. Barley, 1985, "Cultural Organization: Fragments of a Theory", in P. J. Frost et al. (eds.), *Organizational Culture*, Sage, London.

Vickery, G., and G. Wurzburg, 1992, "Intangible Investment Missing Pieces in the Productivity Puzzle", *OECD Observer*, vol. 178 (October/November), pp. 12–16.

FUTURE USES OF CELLULAR AND MOBILE COMMUNICATIONS
YALE M. BRAUNSTEIN

1. INTRODUCTION

The growth of mobile and portable communications in the Asia-Pacific region can be traced to the introduction of cellular systems over the past ten years. Cellular telephone technology enables more efficient use of the radio frequency spectrum as a particular frequency can be used in several (non-adjacent) "cells." This increase in the number of calls that could be served simultaneously led to the increased demand for mobile and portable cellular telephones, both in the region and worldwide, and a drop in the per unit price of the telephones as output increased.

In the future, we shall see both a continuation of current trends and the introduction of new products, services and uses. This chapter starts by describing certain aspects of the current situation in order to help us understand how we arrived at the present patterns in cellular penetration and use. It then moves to a consideration of the near-term future (less than five years), and a discussion of longer-term prospects (over five years). It concludes by making a few connections between what is happening in mobile communications with the changes in multimedia and other advanced communications technologies.

There are a variety of mobile (vehicle-based) and portable telecommunications technologies. These include the older trunked mobile systems, radio dispatch services, and the current cellular radio-telephone services. Cellular systems generally use frequencies in the 450 and 900 MHz bands and involve the "hand-off" of a call from one land-based transmitting and receiving site to another as the user moves from cell to cell. In this chapter, I shall use the term "cellular" to refer to these telecommunications services and the telephone sets using these services regardless of whether the use is mobile or portable.

2.   REVIEW OF THE CURRENT SITUATION

2.1. Major Uses and Users

In most countries the largest group of users of cellular telephony consists of corporate users, smaller businesses, and professionals. They primarily use cellular telephones as ancillary to the established land-line service. Cellular telephones enable these users to obtain functions not provided in traditional, wired-in-place telephone sets. Telecommunications opportunities are expanded by time of day and location. The choice between a mobile or portable unit often depends on traditional work and commute patterns.

Another group of users seeks alternate service providers for financial reasons. Depending on the system architecture, they may be able to communicate with other cellular users or parties outside their local service area without using existing local wireline facilities. This "bypass" of the traditional network may be "economic" in that the cost of providing cellular service is below that of the wireline service or may be "non-economic" which is described by the situation where the true cost (in resources) of cellular is higher but tariffs or other regulatory conditions result in cellular being priced lower to the user.

Cellular and other mobile services are often able to operate even if the wireline service is temporarily out. For this reason they are useful for emergency telecommunications and disaster relief. Similarly, any of a variety of mobile telecommunications services may provide communications to remote areas not served by the wired systems, and several Asian-Pacific countries are exploring cellular, satellite, and other technologies to extend or upgrade their telephone networks.

Although pre-cellular mobile telecommunications systems could and did provide these services, the introduction of cellular systems, with their increased capacity, has led to a significant increase in the use of mobile telecommunications in the Asia-Pacific region as in many parts of the world.

2.2. Remaining Problems

Existing analog cellular systems have inherent limitations on their ability to use bandwidth efficiently and in terms of their coverage. Along with voice channels, additional channels are necessary to carry control information. The frequencies already allocated are becoming saturated in some areas.

Furthermore, in some areas, coverage problems arise from the lack of additional frequencies available for new cells that are not completely isolated from existing cells. There is also a lack of standardization worldwide. In fact, the problem is the existence of too many standards. Currently, there are two frequency ranges, 450 MHz and 900 MHz, and a variety of incompatible analog transmission standards. This leads to less than optimal production runs of necessary chip sets as well as redundant development costs both for subscriber equipment and for central office and cell-site hardware and software. In addition, there are restrictive sales practices in certain countries that result in high service costs, high costs of the subscriber, both.

There are also a number of non-economic factors that have kept cellular service demand from growing as fast as one might expect. There is a (perceived) lack of privacy and security as conversations conducted through cellular phones are broadcast "in the clear" (i.e., are not scrambled or encrypted) and can be monitored by anyone with a suitable receiver. Originally, it was believed that the normal changes in frequency as one traveled from cell to cell would provide a useful degree of security. However, we now realize that two factors have reduced the degree to which one's conversations are truly private. First, the popularity of portable (as opposed to mobile) cellular telephones means many users never change cells during their conversations. Second, the widespread availability of highquality UHF scanners permits those seeking to listen to cellular conversations to do so with relative ease.

The last non-economic factor is related to the perception of a possible health issue. Again, it is the growth in the use of portable units that is forcing the industry to reconsider this issue.

3.   THE NEAR-TERM FUTURE (CIRCA 1995)

3.1. The Conversion to (a) Digital Standard

In some areas, the original decisions concerning standards were made before anyone had knowledge of the usage patterns that would emerge. Today, governments and industry bodies can benefit from their knowledge of who uses mobile and cellular systems and where the usage occurs to choose standards and architectures that allow for more efficient use of resources, for regional cooperation, and for cross-boarder roaming.

The conversion to digital standards will, in itself, increase the effective bandwidth available to mobile communications. It may also make multistandard sets more likely and will provide improved security. However, the new digital sets will likely cost more, particularly at the start, than the analog sets they will replace. It is possible that dual-mode (analog and digital) units will be available during the transition, but they will also carry a cost premium.

3.2. More Emergency and Disaster Relief Uses

Recent natural disasters have shown public safety and disaster relief agencies how useful widespread cellular service can be. As these and other occasional uses become more widespread, we can expect to see the "banking" of telephone numbers for occasional uses and the adoption of special rate plans to meet this special need.

3.3. Ancillary Services and Integrated Equipment or Services

Whether provided by the network, built-in to the hand-set, or attached by the cellular user, the use of and nature of ancillary services will expand. These services include FAX and data transmission and reception, access to a wide variety of information services, voice store-and-forward services, etc.

The upgrading of the cellular systems to increase efficiency in carrying voice communications will, simultaneously, improve the network's ability to carry FAX and data.

New, lower cost alternatives to cellular communications are already under trial (one example is the Telepoint service in London with its outbound-only service). Likewise, equipment designs and features are being rationalized so that the distinctions between cordless telephones and portable cellular telephones are becoming blurred. Not only are these trends likely to continue, but we should expect to see new hybrids and adaptations of multiple services. There are several possible configurations one might expect. They may be as simple as cellular telephones with built-in pagers for when one is either away from a cell site or in an area where communications by one mode is blocked. Another approach is the integration of cellular communications with "personal digital assistants" or some other form of small computing device.

3.4. More Bypass and Alternate Architectures

As cellular telephone service areas expand, operators will be increasingly faced with the question of which functions should be left to the wireline carrier. When a cellular subscriber originates a call, there is no technical reason why any part of that call must be handled by the local wireline carrier with the singular exception of when the call is placed to a local wireline-only subscriber. (However, there often are regulatory or business reasons for using wireline carriers to connect to competing cellular services, long-distance carriers, etc.) We can expect to see more bypass, both of the economic and non-economic varieties, in the future. This will be driven both by further liberalization of regulatory regimes and by the introduction of a wide variety of new local and long-distance services.

The coming changes in existing services (e.g., analog cellular systems converting to digital technology) and the introduction of completely new services will combine to change the ways in which cellular systems are configured and the points at which they interconnect to other communications carriers. Although the earliest of these changes are mostly predictable, those further out in time may be in quite unexpected directions.

4.   THE LONGER-TERM FUTURE

4.1. Changes in Services

It is quite likely that telecommunications in the Asia-Pacific region, as elsewhere, will undergo dramatic changes in the next five to ten years. For this reason it is probably unwise to simply extend the current trends out to a specific year and use that as a starting point for predictions about the future. While one might have a reasonable knowledge of existing technologies, whether implemented or not, and be somewhat confident of the set of possible technologies and services that will be introduced in the short run, this approach is unlikely to be satisfactory for the longer run as some of the technologies may not even exist in laboratory settings.

Despite this disclaimer, I believe that one can reasonably make certain predictions. The first, and in certain ways the most obvious, is that there are likely to be convergences between and among currently disparate services. Obvious candidates include:

1. mobile and marine communications
2. trunked-mobile dispatch and cellular or satellite communications or both
3. tracking, positioning, and two-way telecommunications

One obvious area for this convergence is in numbering plans. Currently, in most regions certain exchange prefixes or blocks of numbers are assigned to cellular telephones. Often it is possible to forward calls from cellular telephones to wireline numbers and vice versa. Similarly, many systems integrate some form of voice mail with call forwarding. Although these approaches provide an increased level of functionality to the mobile or portable telephone user, they do not provide complete integration of a single "follow-me" number. This plan, in its complete version, gives the telephone subscriber a single-700-series in the United States number for life. No matter where the person moves or travels, and no matter which service is employed,

72

when others call the special number they are connected to the recipient's telephone of choice.

4.2. Changes in Hardware

The discussion so far has mostly focused on services. However, in many ways the future of various pieces of equipment are related to the likely changes in services. For example, as services converge, interesting questions arise concerning the future of devices such as cordless telephones and stand-alone pagers. Will they disappear or will they become ubiquitous? Both sets of scenarios seem equally plausible. One could easily imagine a scenario in which all new wireline sets employ cordless technologies as readily as the scenario where the choice is between a traditional wired telephone and a portable unit with capabilities far beyond that of the current cordless phone.

Given this interaction between services and equipment, it seems likely that mobile communications will see a replay of the current wireline battles over whether (and how much) intelligence and memory should be located in the network and its switches or in equipment located at customer premises. Although it is impossible to know exactly which products and services will become the settings for these battles, possible candidates include voice store-and-forward services, FAX networking, and FAX transmissions with voice annotations.

5.   CONCLUSION

Throughout this chapter I have, for the most part, avoided predictions about which new pieces of hardware ("personal digital assistants," for example) and which new services will be the high growth areas in the future. However uncertain the specific configurations of mobile and portable equipment in the future may be, there are three results that seem reasonably certain:

1. The demand for mobile and portable services will continue to grow in countries with a wide range of per capita incomes. These services and the features they provide are useful to a diverse mix of users in a number of settings.

2. Despite the new functionalities being engineered into small mobile and portable units, it is important to remember that, *at the same time*, capabilities and bandwidths of the older wired, switched networks are being improved. As a result, the "functionality gap" between the traditional services, on one hand, and the mobile and portable services, on the other, will continue.

3. Nevertheless, the mix of new and growing systems and architectures—including cellular, satellite, and other non-wireline services—will continue to put a variety of economic and financial pressures on traditional wireline telephone services. This should no longer be viewed as simply a question of how to deal with the bypass

of the "last mile" to the subscriber. The new systems will offer an increasingly diverse mix of services and will need to rely on the traditional systems to an ever-lessening extent.

MULTIMEDIA, HYPERMEDIA AND TELECOMMUNICATION: SEEING THE SOUNDS

DAN J. WEDEMEYER

Global
L96

1.   INTRODUCTION

Not long ago, my three-year-old niece, Amelia, took my hand and led me outside of my house. As we stood on our lanai (patio), she looked up and pointed. "See the birds," she said. I looked up to find nothing but blue sky and clouds. "There are no birds up there," I said, " where do you see the birds?" Amelia replied, "I see the sounds."

As human beings, we have always communicated in a multisensory setting. Only recently, in the past century and a half, have we been handicapped by the limitations placed upon us by our so-called modern telecommunication systems. These innovations allow us to communicate at a distance, but also at a very high price. The price was, and for the most part still is, that we were compelled to trade off being able to express ourselves in an integrated range of audio, visual, and textual venues.

It is only now that selected, state-of-the-art technologies have re-introduced the possibilities of multimediated, distance communication.

But we are just beginning this exciting renaissance, and we have a lot to learn (relearn) and much to do before telecommunication can serve multimediated communication with ease, full access, and at reasonable costs.

The purposes of this chapter are many. First, it is to clarify what we mean when we speak of multimediated, interactive multimedia, or hypermedia. Second, we will look at some of the possibilities and some of the problems associated with networking multimedia or hypermedia applications. This will include both the technical and legal aspects of these emerging systems. Finally, selected issues will be examined, future possibilities set out, and some general conclusions will be drawn.

2.   DEFINING AND REFINING THE TERMS AND CONCEPTS

Multimedia—the combination of text, audio, and visual elements in a single communication setting—has been part of our communication environment for many centuries. In its purist sense, multimedia is sequential (i.e., linear) and non-interactive. The user cannot shape the message; rather, he or she must experience the different mediums as set out by the sender. Early forms of multimedia involved *mixing* media into a common format or setting, e.g., on television or in the classroom. Often times, the mixing of media into a multimediated format is delivered in a linear and non-interactive manner.

Interactive multimedia, or hypermedia, on the other hand, while multimedia based, is nonlinear and user-controlled. Inherent to interactive multimedia and hypermedia (the terms will be used interchangeably in this article) is user and computer control. So, hypermedia is simply nonlinear, interactive multimedia.

Hypermedia is not a simple extension of these media. Hypermedia represents a wholly new medium of communication. As such, it creates an entirely new communication paradigm and will require entirely original ways of thinking and combining of communication skills/strategies. These inherent requirements will affect the designers of messages, the purveyors, and the users of interactive multimedia.

2.1. Hypermedia Designers

These people are the new artists of our time. More often than not, these modern artists operate in teams, combining skills of visualization, scripting, graphic design, and animation along with advanced computer software skills and content-specific advisors.

The designers, at this early stage of hypermedia development, are confronted with basic problems of creating novel structures and techniques of expression. Just as early film makers had to invent and refine linear structures and techniques to tell their story, hypermedia designers are confronted with multidimensional and multiple-linkage problems of design.

Early film makers also faced a world without benefit of special effects to connote meaning. They had to invent such techniques as the dissolve, the wipe, superimposition, and so forth. Today's hypermedia designers are grappling with expressing themselves without benefit of a well-defined package of techniques; these techniques will evolve. The contrast between the first talking motion picture and the latest Spielberg special effects extravaganza will be at least as different as today's hypermedia products and those commonplace in a decade.

2.2. Hypermedia Purveyors

Interactive multimedia purveyors hold equally many opportunities and challenges as the aforementioned designers. While early hypermedia products are "stand-alone" in nature (such as compact disc television or CDTV), the call for networked interactive multimedia will be soon in coming. When the cry comes, telecommunication carriers will be confronted with the need to provide access to broader-band switched services.

The visual aspects of hypermedia will be bandwidth intensive and far beyond what is envisioned for basic or primary rate ISDN. Even with the much needed advances in compression techniques (discussed later), full-motion, color video will most probably require multiple T1s to supply enough

resolution to users whose tastes have been refined by enhanced televisions or

HDTV. The new purveyors and carriers will also face the challenge of serving

a mass individual audience of consumers and not the so-called homogeneous

or even "narrow-cast" audiences of today.

Broadband ISDN (B-ISDN) is not necessarily the answer. B-ISDN

standards are still in formulation and may not be agreed upon until the latter

part of this decade. In addition, the proposed basic B-ISDN channel

bandwidths, in terms of multiple T1s, may not be widely available or

adequate for widespread hypermedia distribution.

2.3. Hypermedia Users

Because hypermedia is controlled by the individual user and is

interactive, the primary focus in design and delivery is the user. In this

regard, then, hypermedia turns the mass media paradigms upside-down.

That is, instead of mass "viewing" or "receiving," hypermedia users speak of

"self-directed navigation" within a very large and rich information

environment. Multimediated resources are used much like a person who is

dining would order his or her food and wine in a fine restaurant with an

extensive menu. No two hypermedia users, like no two diners, would select

or sequence from the information menu in quite the same way. Navigation is based upon the individual's tastes or needs; it is interactive and user-directed.

Unlike dining in a fine restaurant, the individual hypermedia user may decide that one or more of the items he or she wants resides elsewhere. He or she may exit the local information menu and access distant resources. A likely scenario of a hypermedia user in the future is that some of the nonperishable information would reside locally on high-density storage devices, and other perishable, exclusive, or very expensive multimedia databases would be accessed via telecommunication (perhaps from a "video jukebox"). A mix of access capabilities will be required.

3.   POSSIBILITIES OF HYPERMEDIA

As with all innovations, no one is absolutely sure that they will survive or expand dramatically. However, all trends point to the likelihood that hypermedia will not only survive, but will transform the way we create, store, process, and share information.

3.1. Service Opportunities

The opportunities presented by interactive multimedia are almost infinite. This is because hypermedia is not exactly like any existing information service, yet it can accommodate, indeed incorporate, any or all of them. More than likely, hypermedia will enhance each existing service as it creates a wealth of never-before-offered services.

Two of the more promising functions which hypermedia will likely serve are entertainment and education/training. Hypermediated courseware packages and very sophisticated interactive games are already in the development and marketing stages, with many more to come. In addition, specialized and tailored hypermedia packages may well serve promotional and marketing needs—e.g., a new electronic point of sale (POS) brochure combining multimediated advertising and promotion. In addition, the user could complete the purchase transaction.

In addition to the prepackaged hypermedia offerings, it is likely that "generic" multimediated resource discs, already copyright cleared, will be mastered and marketed to education and training specialists and other entrepreneurial types. Such resources would facilitate the "authoring" of a plethora of hypermedia products. The generic disc offerings would most likely come from existing media giants who already have at their disposal thousands of audio and video holdings. These existing resources may be "repurposed" (i.e., used for a different purpose than they were originally intended) to serve the expanding needs for preproduced, copyright-cleared, high-quality, multimedia resources for hypermedia developers.

3.2. Technological Possibilities

Co-evolving with these new services will, out of necessity, be ever-expanding, high-density storage devices; faster and integrated multimedia processors in desktop computers; advancing compression algorithms; high definition input and output displays; and intelligent user interfaces. Hypermedia user terminals will range from dedicated, simple-to-use devices which attach to a television set (e.g., CDTV) at the low end, to very intelligent and full-range hypermedia desktops. Portable versions, expansions of the "Smart Book" or the DynaBook, will round out the users' technical options.

Delivery of hypermedia materials will take many forms. In the stand-alone form, high-density storage devices, compact disc interactive (CDI) and video discs (in the short term) will serve consumer needs. In the networked hypermedia system, one may see broadcast "packet-video" technologies, wireless cable television, or traditional (but fiber-based) cable or hybrid telephone systems serving the market.

Hybrid (duel) systems may be employed, where interactive hypermedia exchanges may be served by narrowband systems (e.g., the traditional telephone or cellular telephone technologies) to request information from distant multimedia sources, and returned broadband requests would use "addressable" broadcast or wireless cable services employing spread-spectrum techniques to individual users.

While this section by no means covers the range of technological possibilities, it serves to alert present telecommunication and information providers of pending changes and challenges. Some of these will be discussed in the following sections.

4.   HYPERMEDIA PROBLEMS AND ISSUES

Almost anytime there is dynamic growth, as is anticipated in hypermedia and telecommunication, there are also pending problems. To the extent that one can anticipate and address these issues early on, opportunities are gained and efficiencies are afforded. This section highlights some selected problems and issues, and in doing so, will alert astute information and communication professionals to high priority actions.

4.1. Hypermedia Problems

Computer history has shown that users will only employ "stand-alone" technologies for a short time before demanding networking (i.e., communication). This will no doubt be the case with hypermedia.

Creating information not only calls for collaborative work in many cases, but often results in a product which can be marketed in network-accessed form. Further, the economies of scope and scale created by networked hypermedia also argues strongly for building networks capable of handling multimedia applications in the near future.

The video aspects of multimedia make the entire application "bandwidth intensive." This obviously presents a problem for most existing networks. Presently broadcast and cable TV networks have the bandwidth, but lack the switching capabilities; at the same time, existing telephony networks have switching, but not the bandwidth necessary to carry high-quality video applications. Changes should be on the horizon.

The basic and primary ISDN networks can handle only 144 kb per second and 1.5 Mb per second (also referred to as T1), respectively. Without major advances in data compression techniques, these speeds will be unsatisfactory to hypermedia users.

Broadband ISDN (B-ISDN) technologies, still lacking a standards agreement, work in multiples of T1s. As such, they will limit economic and ubiquitous delivery of video to the desktop or to the home.

Full-motion, color video requires approximately 90 Mbs or the equivalent of 60 T1s. This is clearly unacceptable and indicates a need for major advances in data compression. In order to accommodate high-quality, switched video in standard ISDN "B-Channels" (64 kbs), algorithms would have to be able to accomplish about a 1500:1 compression ratio. Such advances, which are not impossible to accomplish, do currently present problems for design engineers. Once these problems are resolved, however, compression ratios of this magnitude would not only make widespread hypermedia applications available on public networks, but they could also serve the "tera-level" storage requirements of multimedia resources.

A second, quite different, approach to the delivery of interactive video may be viable. Hybrid wireless systems that combine cellular telephone and wireless cable may be instituted. In such a scheme, the hypermedia work station could request information utilizing the narrow-band cellular telephone network. If the request involved video, the broadband channel, wireless cable, employing spread-spectrum techniques and addressable converters would deliver the request to the individual hypermedia user. This configuration would promote the development of visual databases and reduce the necessity of storing giga-bytes of information at each user's work station.

These approaches are by no means intended to be exhaustive in solving the networking of the video aspects of hypermedia. They only serve to point out the problem of networking multimedia. Most certainly, other solutions will be proposed and some instituted, but for now, switched, interactive video remains one of the major obstacles to widespread delivery of consumer-priced hypermedia products.

While the bandwidth and switching problems are being addressed, at least two other major obstacles are also in need of immediate attention: intellectual property rights (partially a copyright issue) and the production of high-quality, hypersoftware.

These problems are not independent. Intellectual property rights, including copyright, interfere with hypersoftware development. Few people argue with the notion that the expensive development of video, audio, and text should be protected. Not to protect them means that fewer people or organizations would engage in their costly development. On the other hand, overprotection, or the absence of some mechanism of easy and affordable access to these copyrighted resources, severely hinders the development of hypersoftware.

For each hypersoftware developer to become his or her own producer of copyright-free material is both a creative mistake and a "barrier to entry." The development of a range of high-quality video, audio, and other high-end multimedia resources is expensive and time consuming. "Off-the-shelf" resources are often higher quality and certainly speed and ease hypersoftware development. It is here that resolving copyright and intellectual property rights issues come into play. Seeking the source of quality multimedia resources is often difficult; and once the source is found, the "oneoff" negotiations and "use charges" are expensive. New arrangements must emerge.

What may be required is a mechanism, perhaps an organization or enterprising association, that would act as a resource access and copyright clearinghouse. Hypersoftware developers could employ this service to obtain prearranged approvals for video, film, and sound graphic resources at affordable costs. In addition to the benefits to the hypermedia developers, the suppliers of such media would enjoy a broader market for their products. The hypersoftware developers could get on with the job of producing higher quality and more timely hypermedia applications.

A second service may also be spawned by the needs of hypersoftware developers. Generic or tailored video discs or CDs (like the presently available audio discs) which contain pre-cleared copyrighted materials may be mastered and sold to hypermedia developers for unlimited use. Another solution may involve "dial-up" video jukebox services, where a developer could select and pay for a particular piece of "video clip art" to be used in software development. In such schemes, the multimedia developers would have the "raw resources" from which they could design and build quality products. The result would be more and better products at efficient production levels at consumer-affordable prices.

Once the previously mentioned technical and legal problems are addressed and resolved, the artistic difficulty of producing high-quality hypersoftware still remains. As many are aware, producing any single type of programming (e.g., video) is a complex task. Multimedia software, by definition, magnifies these single medium problems, and as such, its order of magnitude is more difficult. Usually a team approach is employed.

A full team, at a minimum, may consist of a project facilitator/ expediter; a technical specialist; a writer-designer; a video producer; a computer programmer; a content specialist; and a graphics designer. As with any team approach, clear and concise objectives must be set out and closely coordinated/managed. The users' needs must be well understood and central to the design and testing of the hypersoftware. Development is often tedious and sometimes lengthy. Hypersoftware development follows the rules of most media production: it always takes a little longer and costs a little more than expected.

5. THE IMPACT OF HYPERMEDIA

We shape our tools, and, in turn, they reshape us. This may be particularly true with regard to our communication and information tools. As human beings, we have been living in an electronically narrowband and mass-mediated world. Interactive hypermedia will change much of that reality. Markets will be mass, but individual at the same time. People will mass consume, but each user will use the media in an individualized manner. Our electronic world will move one step closer to our multisensory, real world. This obviously has many social, political, and economic implications. In the following sections, selected impacts of our new information tools will be examined.

Networked, interactive, broadband media will change us as individuals and will change the institutions in which we work. The implications for the way we work, the way we play, and the way we learn will be significant.

Networked, "virtual" business organizations and distance-learning networks shift existing power structures, politically and economically. Each of the functions of business will be electronically altered. Trade flows change as a function of our new or modified communication paths. Money follows those same paths, and as a result, politics change. Power shifts.

Education networks will take advantage of distant, higher-quality information resources, both people and electronic. Electronic-based education markets, once viewed as a relatively minor sector, will take on a leading status among telecommunication carriers and information providers. Each will have to take into consideration the advances of interactive multimedia.

Socially and culturally we will also be changed. We will be simultaneously living in two worlds; one local and one global. Our local cultures will evolve and change as a function of our new communication capabilities, and we will develop (are developing) a new global culture based upon information technologies and services. As Toffler, Nesbitt, and others have pointed out, we used to form our interpersonal relationships on the basis of vicinity; now, and in tomorrow's world, we will be forming and maintaining our relationships on the basis of affinity. Slow, according to Toffler, has always given way to fast. The demands of the new global information technology (IT) culture will impact our local culture. IT will speed local culture and the global network will start to dictate local change. This will be primarily due to our emerging communication tools, which are increasingly multimedia-oriented, networked, and, of course, almost instantaneous.

6.   THE NOT-TOO-DISTANT FUTURE

While we have been dealing with the future in the previous sections, it has been the immediate future that was being referred to. Hypermedia is not the end of telecommunication evolution, but rather is the beginning of a new era which will spawn other, more exciting, opportunities. Hypermedia, however, gives us a clue to at least one evolutionary path, and that path will be discussed in the following paragraphs.

Hypermedia, as set out in this chapter, draws many media into a computer and user-controlled environment. As technologies, software, and imagination converge, we will see an extension of these media into what we now know as "virtual reality." Virtual reality takes hypermedia to the extreme—i.e., it utilizes all of the multimedia platforms—and extrapolates them in order to "fool" our senses into believing that what we are experiencing electronically through our senses is, indeed, real.

Virtual reality already exists in such applications as AutoDesk. It allows us to assume a different reality, experience it first hand, and be amused by it, or learn from it. For a short time we can experience what it is like to live as another animal (e.g., a lobster); we can tour the interworkings of a molecule (inside DNA); or we can navigate around, and view first hand, another planet. Through a more total hypermediated environment, we are transported into, and control, a separate reality.

While it will be a while before these virtual realities are available upon demand through our telecommunication networks, I believe that the experiences we have gained in data networking and hypermedia networking will provide insights into the requirements of a not-too-distant future when we can electronically step out of our present reality and experience alternative, virtual "real" experiences.

7.   CONCLUSION

The 1990s will most likely be considered the "visualization decade." It will set the stage for a time when we will be able to easily and affordably access multimedia networks upon demand. If we are successful, the first decade after the turn of the century will be characterized by relatively ubiquitous multimediated networks, many with "virtual reality" connections.

The characteristics of this emerging environment will be broadband, interactive, and nonlinear. The resources drawn upon by the hypermedia user, in many instances, will reside at remote sites, drawn upon, as needed, through the desktop or personal communication gateway located anywhere on the globe.

There are many implications for change brought about by the evolution toward integration of video, voice, and data. Technical innovations will obviously be required and, I believe, are forthcoming. These innovations will call into question other legal, economic, social, and cultural issues that were discussed briefly in this chapter. Finally, there will be many implications for the individual hypermedia user. At a minimum, these new networked capabilities will change the way he or she will work, learn, and play in the decades to come.

In the not-too-distant future, we will be calling for "virtual reality networks" that are capable of supporting even higher bandwidths introduced by HDTV and multiple-user "conferenced" experiences. Hypermedia signals a need for a change in telecommunication. As the tools that we use for communication change, so too will we change. For better or for worse, we will be living in a hypermediated, increasingly "virtual" world. In such a world, adults, along with three-year-old children, will be able to "see the sounds."

CULTURAL BASIS OF TELECOMMUNICATION SYSTEMS: AN INTRODUCTION

SYED A. RAHIM

The human actors and stakeholders of telecommunications—the manufacturers, vendors, operators, and users—constitute a particular kind of social order. It stands in relation to the technological order of information-communication machines and infrastructures. For example, a telephone conversation is a social event that directly and indirectly involves not only the two persons in actual conversation but many others engaged at that moment in operating, monitoring, and maintaining the system. The communicators, telephone operators, maintenance workers, supervisors, planners, engineers, accountants, manufacturers, and distributors of telephone goods and services are working in a system that makes them interdependent, even if many of them may not have ever met or talked to one another. In general, a telecommunication system is a socio-technical system, a composite of a social system and a technical system. Its social order is based on various social relations which are organized into distinct processes and structures— economic, political, cultural, professional, personal, and others. The telecommunications system is built on these structures. Its dynamics is generated by the interplay of those structural bases.

In understanding how telecommunication systems develop and change, it is important to recognize their multiple bases of process and structure—technical, economic, political, social, and cultural. Which one of these bases is most important is not a very useful question, because different bases may play different and variable roles and functions depending on the specific purposes, the stakeholders involved, and the historical, spatial, and temporal contexts of telecommunication. But one can validly focus attention on a particular structure assuming that it is an essential and important component of a larger system. In this paper, I focus on the cultural structure as one of the bases of telecommunication systems. The objective is to throw some light on how this structure is evolving along with other structural bases of modern telecommunication systems.

1.   CULTURE: SHARED AND CONTESTED

The participants of a telecommunication system (or any other social system) in society may come from different economic classes; political persuasions; social, demographic, ethnic or professional groups; and spatial communities. They bring with them their particular ideas, world views, values, professional knowledge, language, and motivations. These are some of the ingredients they use in constructing a subculture; the technology of communication supply other ingredients. The nature of technology and innovations offers opportunities and sets constraints in the process of developing a structure of telecommunication culture. The process of cultural formation is not a simple aggregation of values and beliefs but involves competition, conflict, disagreement, and disappointment. The participants remain engaged in a continuous struggle, as the technology, economics, and politics of communication develop and change.

The culture of telecommunication may be defined as an organized set of a particular kind of nonmaterial social elements or social facts about and relevant to telecommunication systems. These social elements or facts are both shared and contested by the members of the telecommunication community. The structure and organization of the elements, and their relationship to one another and to outside elements, influence telecommunication systems in different ways.[1]

General examples of such elements for any culture are ideas, values, norms, beliefs, tastes, etiquette, symbols, and meanings. These elements belong to the stakeholders' individual and collective consciousness and they are primarily social in character. All the members of a community or a subgroup may share some of the elements, and different subgroups may contest the relevance and significance of some other elements. For example, in the United States, most people share the idea of "the freedom of speech" as a central sociopolitical value. But "freedom of abortion" as a value is highly controversial; two large groups of people have opposing ideas and beliefs on the subject, yet both are part of the contemporary American culture. In practical applications, a widely shared core value, like freedom of speech, is subject to specific interpretations in different contexts. This creates an opportunity for stakeholders to contest each others' interpretations to protect their interests or to represent their positions when controversial issues are raised and debated. Thus sharing and contesting are two aspects of the same internal dynamics of cultural stability and change.

The culture elements of telecommunication are represented in various material products and events of human communication activities. Examples are: consumer or industrial product designs and the way those are advertised, commercial and scientific exhibitions, and telecommunication protocols. But the culture elements themselves are not material objects or behavior. The elements (such as values or tastes) have a distinct reality of their own. Also, culture is not just an aggregate of such elements but is a meaningful organization of such elements. The internal relationship among the elements constitute a structure of culture. This structure and its connections with other structures (economic, political, social, and other cultures) are the means through which culture interacts with and influences society at large.

The social nature of the telecommunication culture makes it a public rather than a private phenomenon, except in the case where private means

99

exclusive ownership or restricted access. Telecommunication culture as a distinct social reality is related to but exists apart from physical or psychological aspects of communication and human interaction. It is a social resource or capital which can be accumulated, mobilized, and used to bring development and change in society. This power of telecommunication culture is not equally shared by all members of the community. The stakeholders as individuals and groups compete with one another for cultural capital and power. The power of cultural capital tends to be associated with the power of economic and political capital, and is concentrated in the public and corporate policymaking bodies of society.

2.   NATIONAL TELECOMMUNICATION CULTURES

National telecommunication cultures significantly differ from one another, even though many of them may share some core elements. In the United States, the basic elements of telecommunication culture are derived from a set of core values, ideas, beliefs, norms, and standards: the freedom of speech, protection of telecommunication messages from censorship and privacy violation, universal and equitable access to basic communications services, free market communication economy, individual choice and responsibility in seeking and judging communication contents, diversity and efficiency of services, government regulation for serving and protecting public interests, federal government control of communications for national defense and security, and communications technology as the cutting edge of capitalistic and democratic development of the nation. An evolving organization of these and other derived elements—some mutually reinforcing and others in a relationship of tension—have influenced the growth and development of telecommunication policies, industries, and services in the United States. As Roger G. Noll pointed out, telecommunications policy evolved in a climate of value conflicts between free market and government regulation, inexpensive basic local services and advanced services for increasing productivity, and efficiency through integrated firms and competition under market and economic regulation.[2] In that process, the telecommunication culture itself has changed and its main premises have been rearticulated in diverse contexts.

A cultural study of the historical development and change of telecommunication in the United States will be very useful in understanding the trends of the future of national and global telecommunication systems. Such a study may involve how cultural elements are interpreted, combined and used, focusing on such main values and events as the First Amendment, the 1935 Communications Act, establishment of the Federal Communications Commission, divestiture of AT&T, and the regulatory contests in broadcasting, cable television, satellite, and computer communications. A study that examines how various cultural elements in different combinations create different kinds of tension and produce different results in the development of telecommunications as a whole as well as its different

components—including the common carrier, broadcasting, mobile

communications, satellite and computer communication networks—could

provide great insights. However, such a study is beyond the scope of this

paper. But the point is raised here to underscore the importance of the

cultural basis of modern telecommunication systems.

In Japan, universal basic services, freedom and privacy of

communication, a perceived rapid trend of domestic and international

informatization of society, belief in technology-driven progress, and the value

of telecommunications in providing support to active participation of a rapidly

increasing older population are important elements of the telecommunication

culture. Traditionally, the value of a consensual process among three

powerful institutional actors has been central to Japanese telecommunication

culture. These actors are the Ministry of International Trade and Industry

(MITI), the ruling Liberal Democratic Party, and Keidanran (the Federation of

Economic Organizations). At the same time, competition and conflict among

those actors have been driving forces of recent changes in the

telecommunications policy and industry. Like the United States, Japan has

taken bold actions of privatization and liberalization in the arena of

telecommunications. Yet the power of the trio (Japan Inc.) has remained

supreme (Ueda 1990, Hills 1986, and Ministry of Posts and Tele-

communications 1990).

In the Federal Republic of Germany, the government assumes the ultimate authority and responsibility of providing a wide range of communication services through the marketplace. There is an emerging belief that regulatory reforms can and must promote innovation and competition and guide the private sector in providing a variety of communication services, while the government telecom agencies should continue to control certain basic services so that universal access and equity can be ensured (Federal Ministry of Posts and Telecommunications 1988). In France, President Mitterrand's call for democratizing computers instead of computerizing society shows how contending capitalistic and socialistic ideological values dominate contemporary French telecommunication cultural discourse (Matterlart and Cesta 1985). In Singapore, the ruling party leaders value modern communication media as a powerful means of administrative control, education and public information, and opinion formation. Hence it is essential to formulate public policy for appropriate control and guidance of the media. Adequate control and supervision is necessary to ensure that the media is not manipulated against national interests, but used for nation building and social development (Kuo and Chen 1983).

3.   CULTURAL AND STRUCTURAL CHANGE

Since 1980 major structural changes have taken place in the telecommunication systems of leading industrial countries. In the United

States the change has been dramatic. It has produced a significant demonstration effect and, in some cases, political pressure to reform telecommunications in other countries. Within a decade, the arena of telecommunications in the United States changed from a regulated monopoly of AT&T geared to affordable universal service to a battlefield of competitive industries offering a wide range of communication services and products at different price levels to all kinds of national and international consumers. The divestiture of AT&T in January 1984, and associated decisions by the Federal Communications Commission (FCC) before and after that event, established a new regulatory regime geared to liberalization and privatization of telecommunications. These extraordinary set of events generated a great deal of interest and concern in telecommunications industries and businesses all around the world. Japan and the United Kingdom promptly took comparative action to make their telecommunication systems competitive.

In Japan, the AT&T's equivalent was the Nippon Telephone and Telegraph (NTT), a giant public corporation under the supervision of the Ministry of Posts and Telecommunications (MPT). A movement for liberalization and privatization of telecommunications evoked intense contest and conflict among important stakeholders of telecommunications, in particular, between the Ministry of Posts and Telecommunications (MPT) and the Ministry of International Trade and Industry (MITI). In 1984, the Telecommunication Business Law and the NTT Corporation Law passed in the Japanese Diet and created the legal framework for structural

reorganization of telecommunications. Under the corporation law, NTT was

privatized. The regulatory reform retained MPT's power to regulate rates. The

breakdown of NTT, a hotly debated issue, continues to be one of the main

items on Japan's telecommunications reform agenda.

In the United Kingdom, telecommunications liberalization started in

1981 when British Telecom (BT), a public corporation, was separated from the

British Post Office, and later when a competitor, Mercury plc, was given a

license to operate specific services. In 1984, British Telecom was privatized

and a 25-year license given to it made specific provisions to prevent

anti-competitive action by BT. The regulatory framework of telecommunica-

tions in the United Kingdom has been redesigned for partial regulation, to

promote unregulated competition where possible and to have effective

regulation where necessary, by segmenting the markets according to their

particular technical and economic characteristics.

The structural changes in telecommunications sketched above are the

result of complex interactions among technological, economic, political, and

cultural factors. Technological innovation transformed telecommunications

from a basic utility to a range of attractive producer and consumer

commodities and services. Declining prices of certain products and services,

and enhanced services for special consumer groups, expanded and segmented the marketplace. The new conservatives in power (Reagan and Thacher) promoted deregulation, liberalization, and privatization policies. All of this happened along with a significant cultural transformation. In the remaining part of this paper, I briefly discuss the broad nature of change in the cultural basis and its future direction.

There are a number of significant shifts along the belief-values dimension of telecommunication structures. The notion of telecommunication as an essential public utility, a single infrastructure of mediated communication which produced and dominated the regulated monopoly structure, is now only one of the major aspects of the telecommunications complex. The concept has expanded to include other meanings of telecommunications: novel and diverse commodities meeting a wide variety of needs and tastes; highly competitive future-oriented industries; multimedia infrastructures; and dynamic field of innovation, obsolescence, and risk. Daniel Bell's characterization of information transmission and processing through computers and telecommunications networks as the transforming resource of the post-industrial society has caught the imagination of many telecommunications stakeholders (Bell 1981). The belief that telecommunications is a natural monopoly has lost its rationality. The rising importance of telecommunications in international trade, political communication, and cultural production has promoted new organized groups of powerful stakeholders fighting for structural change: limited and selective regulation, widespread privatization, multimedia infrastructure, and globalization. The transition from a regulated monopoly to managed competition in telecommunication has been possible because values, beliefs, and meanings of telecommunications have changed.

4.   FUTURE DIRECTION

The impact of telecommunications in society depends on when, how, by who, and for whom technology is used. This, in turn, is regulated by the economic, political, and cultural structures of telecommunication. Each one of these structures is subject to influence by the larger national and global economic, political, and cultural patterns and changes in those patterns.

A modest account of the future can be based on what is already known and being tried or tested experimentally or used on a limited scale. We know that modern computer and telecommunications technologies are used to design virtual classrooms, libraries, corporations, war simulations or world series games (Pimentel and Teixeira 1993, Saunders 1993, Davidow and Malone 1992, and Baudrillard 1988). In the near future, increasing technological capability will make it feasible to use high-quality information and super-real image construction at reasonable costs. Then, such virtual activities in education, entertainment, business, and international relations will significantly alter and complement the existing modes of activities in those arenas. For example, it may be possible to settle an international conflict by fighting a virtual war that will show the consequences most accurately and vividly. The possible outcomes of war experienced virtually may provide incentives and motivation for a negotiated settlement of the conflict.

From a cultural point of view, the most exciting thing is the evolution of telecommunications networking and its extension beyond the conventional voice telephone networking. The convergence and integration of voice, video, and data transmitting and processing in local and wide area networks will have a revolutionary impact on the formation, operation, and integration of social groups and communities in every sphere of life. It will demand more active participation from the individual. More participation at less effort and cost will be possible because of many channels of virtual participation. Attending town meetings or college classes through two-way television, casting a vote electronically, doing office work at a home computer terminal, and many other new forms of work and interaction will become common features of life in the advanced telecommunications society.

A cultural phenomenon of utmost significance is evolving in the United States and other post-industrial societies. It is the rise of symbolic and informational transactions through telecommunications media and its positioning at the leading edge of many spheres of organized activities, including military, trade, politics, community, religion, news, and entertainment. For example, in today's banking and trade much of the business activities consist of symbolic transactions, electronic transfer of money, services, and goods through computers and telecommunications networks. Business deals are completed and profits made virtually before (sometimes without) actual physical transactions of goods and services taking place. Television transforms viewers as virtual observers of real-time events that are electronically imaged on the TV screen at home. An extreme manifestation of this in high-technology culture is a world of self-referential signs driven by human desire and imagination. This is a world constructed by metaphor, metonymy, commodity as information, and simulation. In Baudrillard's term, this is the world of hyperreality (Baudrillard 1992).

The satellite telecommunications technology linked with television technology has made it possible to virtually view and observe events in different parts of the world on a real-time basis. This has become an everyday experience for people watching television. It has radically changed the cultural horizon and world view of people in many different cultures, bringing them closer to one another. As the virtual telecommunication systems develop and expand, the status of the communicators will change from that of a mere passive observer to active participant. Then it will bring another radical change in the cultures of mainly the advanced post-industrial societies.

On the other hand, a very large proportion of the world population lacks access to basic telecommunications services. For those people, the ideas of universal access at reasonable costs and telecommunications as a basic utility are very important. Can advanced telecommunications technology solve the problem of worldwide universal access? Are there cultural factors responsible for the slow growth of telecommunications in many developing countries or is it primarily an economic problem? How will the growing imbalance in telecommunications access and use affect cultural, economic, and political exchange between industrial and developing nations? These questions are important and need economic, political, and cultural analysis to find useful answers.

Recently considerable progress has been made in the economic analysis of telecommunication, but cultural analysis is lagging behind mainly because of a lack of attention and interest from telecommunication scholars and researchers. The telecommunication field needs an integrated research perspective and policy development, as much as it needs technology integration.

NOTES

1. This definition of telecommunication culture is based on ideas on the nature of culture that have been discussed by a number of scholars. See, for example, Bell (1976), Berger (1967), Bourdieu (1977), Douglas and Wildavsky (1982), Durkheim (1964), Geertz (1973), Habermas (1971), and Foucault (1980).

2. Also see Edger and Rahim (1983).

REFERENCES

Baudrillard, Jean, 1988, Selected Writings, Mark Poster (ed.), Stanford University Press, Stanford.

Bell, Daniel, 1976, The Cultural Contradictions of Capitalism, Basic Books, New York.

——, 1981, "The Social Framework of the Information Society", in Tom Forester (ed.), The Microelectronic Revolution, MIT Press, Cambridge.

Berger, Peter L., 1967, The Sacred Canopy, Doubleday, Garden City.

Bourdieu, Pierre, 1977, Outline of a Theory of Practice, Cambridge University Press, New York.

Davidow, William, H., and Michael S. Malone, 1992, The Virtual Corporation: Structuring and Revitalizing the Corporation for the 21st Century, Harper-Collins, New York.

Douglas, Mary, and Aaron Wildavsky, 1982, Risk and Culture, University of California Press, Berkeley.

Durkheim, Emile, 1964, The Rules of Sociological Method, Free Press, New York.

Edger, Patricia, and Syed A. Rahim, 1983, Communication Policy in Developing Countries, Kegan Paul International, London.

Federal Ministry of Posts and Telecommunications, 1988, Reform of the Postal and Telecommunications System, R.V. Decker Verlag, G. Schenck, Heidelberg.

Foucault, Michel, 1980, Power/Knowledge, Pantheon, New York.

Geertz, Clifford, 1973, The Interpretation of Culture, Basic Books, New York.

Habermas, Jurgen, 1971, Knowledge and Human Interests, Beacon, Boston.

Hills, Jill, 1986, Deregulating Telecoms, Frances Pinters, London.

Kuo C. Y., and Chen S. J., 1983, Communication Policy and Planning in Singapore, Kegan Paul International, London.

Matterlart, Armand, and Yves S. Cesta, 1985, Technology, Culture and Communication, North Holland, Amsterdam.

Ministry of Posts and Telecommunications, 1990, 1990 Communications in Japan, Ministry of Posts and Telecommunications, Tokyo.

Noll, Roger G., 1989, "Telecommunications Regulations in the 1990s", in Paula R. Newberg (ed.), New Directions in Telecommunications Policy, Duke University Press, Durham.

Pimentel, Ken, and Kevin Teixeira, 1993, Virtual Reality: Through the New Looking Glass, Intel/Windcrest, New York.

Saunders, Laverna M. (ed.), 1993, The Virtual Library: Visions and Realities, Meckler, Westport.

Ueda, Masano, 1990, Deregulation of the Telecommunications Industry in Japan: Was the Breakup of NTT a Solution?, Program on U.S. Japan Relation, Harvard University.

FIGURING ELECTRONIC MONEY: INFORMATION ECONOMIES AND CYBERSPACE POLITICS

ANTHONY PENNINGS

1.   WHEN IT CHANGED

In the early 1980s, Walter Wriston, then CEO of Citicorp, proposed that the world's financial markets work on a new "information standard," leaving behind the gold-dollar standard created at the Bretton Woods Conference in 1944. The world of international finance and currency exchange trading had been slowly changing from a small club of major bankers to a global network of telegraph- and telex-linked financial dealers. Reuters was not unfamiliar to commerce in financial information when the world went off the gold standard, it had gotten its start with carrier pigeons in the mid-nineteenth century. What was new was the new utility of telecommunications services. When the governments of the major currencies of the world decided to let the "markets" decide the exchange rates, a type of electronic interchange interceded to coordinate these transactions. Instead of being controlled by a few of the largest central banks, the world's currencies became linked together in electronic

116

markets using value-added telecommunications networks and computer monitors displaying the prices of various currencies.

In his nationally televised speech of August 15, 1971, Nixon introduced his New Economic Policy (NEP) which deregulated the dollar and destroyed the gold-dollar standard. The "Nixon Shokku" as it was called by the Japanese effectively removed gold from the international monetary system. It delinked the dollar from its $35 dollar per ounce of gold obligation and thus destabilized most of the world's currencies, which under the Bretton Woods agreement were generally required to be held within 10 percent of the dollar's par value. In the advent of the "Nixon Shock," the electronic environment became the new "market" for buying and selling international monies. Geostationary satellites joined the longstanding networks of undersea communications cables to provide data and voice services to financial institutions all over the world. The same rockets which propelled the first moon landings also provided the launch capabilities for the international telecommunications satellite consortium, INTELSAT, which provided dial-up and leased lines for currency and stock price services such as Reuters' Money Monitor Rates (Pennings 1986). By the time the International Monetary Fund (IMF) had endorsed floating exchange rates in 1973, the Reuters news agency became the central source of prices in the currency markets. "The world since that time has been operating with a monetary system for which there has been no historical precedent in that no major currency in the world is currently tied to a physical commodity" (Wriston 1992, p. 58).

The impulses associated with sovereignty and exchange are intimately tied to the politics of domains, subjectivities, and territories. This makes them a suitable mode of interpretation for the interrogation of electronic spaces in the modern world. Sometimes contradictory, but sometimes mutually cooperative, these signifying surges nonetheless represent fundamental tensions or ebbs and flows to be considered in the analysis of sovereignty and the understanding of the notion of general equivalence as it relates to practices of signification. This chapter considers Wriston's "information standard" in relation to Jacques Derrida's concern with "logocentrism." Using a more interpretative approach focused on the manifestation of symbolic economies, the elevation of "information" is treated as a type of computer logocracy, a variation in the mode of phonocentric signifying which has, with the alphanumerical mode of representation, come to be the symbolizing other of modern signification.

In the age of CNN and the dissolution of the USSR, the notion that a technology-led democracy is breaking out has became very popular. Wriston's view that the state-centric model of geopolitics is dissolving into a sea of democratic proclivities has some interesting discernments into the processes of exchange, but what is disturbing is that he appears to be incognizant of the alternative forms of boundaries and identities that are creating new forms of authority and control. For instance, if we know anything about the consequences of information societies so far, it is that most of them have become debt societies. The new communication and information technologies have merged neatly with the deregulated environment of banking and other financial institutions allowing new forms of computerized debt to register on national, corporate, and individual balance sheets. The new mechanization of these long-time information industries has dramatically changed these organizations. From credit cards to junk bonds and national treasury notes, the new instruments of financial credit and obligation have created a global grid of obligation and engendered a new politics of debt. What started as the "third world debt crisis" has turned into a global debt epidemic with as yet largely unexplored consequences for nations, enterprises, and individual citizens. In the 1970s and 1980s, major bankers lent hundreds of billions of electronic petrodollars, often called eurocurrency, to sovereign governments because they claimed that these governments could not go bankrupt.[1] Heavily burdened with foreign debt, these countries have been forced to undergo IMF restructuring of government administration, programs, and social services in order to procure more funding to pay off the debts. Next came the reduction of wages while simultaneously there was a lifting of government controls on prices. Finally they are "privatizing" their indigenous resources at receivership prices and, with computerized debt for equity swaps, they are seeing wholesale stocks of resources switch from public to private ownership. The new circulation of wealth due to the emergence of electronic money has made it possible to conscript formerly public-owned resources under the codes of private ownership.

2.   PRIVATIZATION AND THE GLOBALIZATION OF INFORMATION

A striking signature of the 1980s will certainly be the dramatic changes in the telecommunications sectors around the globe. Throughout the world, most formerly government-controlled PTTs (postal, telephone, and telegraph) have undergone various degrees of deregulation, liberalization, and privatization. Although these terms are often used interchangeably, their generally assigned meanings within the context of official telecommunications policy refer respectively to the easing of government rules and stipulations; opening up of new markets for telecommunications; and the selling of government or publicly owned facilities to private interests.

At the global level, privatization requires the consideration of at least two dynamics which should be considered as contributing to this massive organizational restructuring. The first is the elevation of "information" as the central organizing strategy, the new currency, for the transnationalized political economy. Second is the proliferation of telecommunications and information technology providers. By approaching the privatization issue from the perspective of both information and information technology, it is possible to see that the interconnections between the two are complex and yet substantial enough to warrant further elaboration.

Combined with the computer, information becomes even more significant. In fact, the second dynamic to be considered in the privatization of telecommunications is the range of new information technology provides introducing network, desktop, and personal technologies at an unprecedented rate. The combination of the new forms of information techniques with the information technologies improve command over the logistical and financial aspects of the modern organization which has been convinced that it needs the new technologies to compete in the national and world markets. They want the most sophisticated technologies to ensure the utmost in coordination and control for the management of transnational operations.

The number of entrants into the global telecommunications and information technology business increased dramatically in the 1980s. Telecom equipment suppliers, many who were previously wed to domestic

PTTs, saw the international market as ripe for their network technologies. NEC, Okidata, Siemens, and others (many with solid government backing) all began to look outward to sell their equipment and services. AT&T was released from the FCC Consent Decree which had limited it to domestic sales, and subsequently, the Regional Bell Operating Companies (RBOCs) or "Baby Bells" also began to look internationally. These companies have been developing an extensive array of new broadband transmission systems and digital switching modules capable of extending their value-added service offerings and have targeted foreign countries as their new markets.

Private firms such as those in the banking, manufacturing, and travel industries have pushed for PTTs and telecom providers around the world to modernize their facilities in line with innovations in global technologies and standards. The considerable investments required to modernize switching systems, trunk lines, as well as "the last loop," have made PTTs reluctant to spend the funds required, especially when it was looked on as a major source of employment and a boost for national treasuries. However, airline reservation networks, currency quoting and trading systems, as well as the new industrial data communication applications over international valued-added networks brought early pressure on PTTs around the world. A number of technological developments such as faster microprocessors, multiplexed packet-switched circuits, high-speed data transmission, digital switching systems, enhanced RAM/ROM memories, as well as elaborate relational software programs, all required new strategical approaches to the business of telecommunications.

The firms that were able to comprehend this trend and integrate the emerging information technologies "leaped into the world arena." To do this, the firms needed to learn from each other by watching which organizational devices and strategies worked and which did not. They had to cognate and master this information by sharing upper management and personnel who understood the "languages" of commercial, financial, and productive knowledge (Dordick and Neubauer 1985).

The identification of information and information technology as a central phenomenon in geopolitical and geoeconomic development is not particularly unique to this study. A whole body of knowledge focusing on the arrival of an information society has circulated in academia and in popular writings for some time now. What portends to be unique in this project is a more extensive interrogation of the term "information" and an analysis which figures this choice of a signifying process as an emerging exclusive form of the knowledge system.

3. WRITING ECONOMIES/DECONSTRUCTING INFORMATION

Up until this point the implication of language and writing technology as a general equivalent has been only briefly mentioned. It is, however, one of the most intriguing of Jean-Joseph Goux's theoretical propositions in his seminal book, *Symbolic Economies*. The ascension of writing is part of Goux's fourfold effort to bring attention to the notion of the general equivalent and is a crucial element in this analysis of cyberspace. His strategy links the modes of writing which were characteristic of different social formations to the conditions of economic and cultural exchange in each of those societies. Before it became fixed in its present phonographic stage, writing moved through historical "thresholds" where it performed as pictographic and then ideographic modes of signifying.

As writing moved into the phonographic stage, there developed a much more fluid and effective way of dispersing forms of knowledge. "Through this general equivalent—the new system of alphabetization—it became possible to translate concepts from one language to another, and authors began to participate extensively in the exchange of knowledge" (Shapiro 1985).

Goux includes the elevation of a "logocentric" signifying practice in the realm of the general equivalent. In his terms, the prevailing mode of notation is congruous with the dominant form of exchange.

4.   READING THE INFORMATION SOCIETY

The technocratic delineations of knowledge achieved a high rate of circulation by the 1980s when the notion of an information society was popularized. In Japan, the concept of *johoka shakai* or "informationalized society" had circulated since the early 1960s when Tadeo Umesao published an influential article entitled "*Joho Sangyo Ron*" (On information societies). Using a biological metaphor, he proposed three stages of human industry development leading to a "spiritual industry" of knowledge production and consumption. The earlier stages of agriculture and material industries were likened to the body's needs for food and shelter (Ito 1981, Masuda 1981).

Marc Porat's doctoral dissertation which disaggregated U.S. Department of Commerce employment data has often been cited as empirical proof that agriculture and manufacturing industries were on the decline and that the "information economy" had arrived (Porat 1977). Activities that created information for sale or as an intermediary enhancing the production process were growing at the relative expense of activities which created farm and factory products. By focusing on the activities that workers performed rather than their explicit job titles, he was able to argue that by 1980, "more people in the United States were engaged in information work than any other kind of work, indeed about 48 percent of the U.S. population was engaged in one form or another of information work, while only about 3 percent were in agriculture, slightly more than 20 percent in manufacturing, and about 30 percent in providing services." (Dordick 1986, p. 9)

For the less enthusiastic Marxist-informed interpretations, "information" is usually explained as a solution to aid capitalism's desire to extract surplus value without extensive investments in additional equipment, labor, and raw materials. The dependency schools focused on how international flows of information extend domination of existing power structures, making weaker states more dependent on the centers of innovation and capital. When not tied to the productivist mode of accumulation, information is usually regarded as part of the realm of the superstructure, following a general epistemological differentiation between reality and appearance.[2]

Information is not a new word flowing through the English language; nevertheless, it achieves a new importance with the technical innovations of the second world war. Its new carnation has strong roots in the military and industrial efforts leading to major advances in ballistics, computing, and telecommunications. Automatic gun sights, radar, and other tracking technology sparked the search for the alphanumerical calculation of electronic signals. Initially developed to target German airplanes for anti-aircraft guns, cybernetics was soon elevated "to attack the problem of control and communication in general" (Weiner 1954, p. 17). As a by-product of the sensory and calculating equipment, "information" emerged as a mathematical and measurable concept. The twin theories of information theory and cybernetics led to the development of a formalized unit of "feedback" and measurement.

Before the end of the 1940s, an executive of the Rockefeller Foundation joined a Bell Lab scientist to produce a book entitled, *The Mathematical Theory of Communication*. The executive applied a more generalized interpretation to the scientist's theory. According to these authors, Claude Shannon and Earl Weaver, information becomes a statistical relation between communication and noise. That which is not lost in the communication process by entropic dissipation is information.

5. PRIVATIZATION AND THE GLOBALIZATION INFORMATION

The Shannon-Weaver model (Shannon and Weaver 1949) outlined in *The Mathematical Theory of Communication* was instrumental in

decontextualizing information from its cultural and political environments. "Systems or information theory is, in a very real sense, the
'ideology,' the metaphysical justification and practical belief system of the
'information revolution" (Finlay 1987, p. 163). The information theory
model assumes similarly enculturated sovereign identities on both ends of
the communication channel and shifts the focus of inquiry to the
problems occurring along the route from sender to receiver.

Unlike textual approaches which acknowledge the complexities of
language and other forms of mediation involved in the exchange of
meaning, the "bullet theory" of communication posits problems in terms of
"interference" or "noise" with an unmediated message. While it does
consider that the sender and receiver respectively "code" and "decode" the
message, it does not consider the implications of signification for
subjectivity. The bullet theory works to reify the illusion that
communicating subjects do not have to enter into a prescribed social
order to be able to exchange meanings. While textualists are quick to
point out that a price must be paid to enter into a discourse, the
Shannon-Weaver model configures the process in technical terms,
abstracting and refining information to render it more susceptible to
forms of exchange based on a limited mode of signification and identity
coherence.

According to Goux, the form of writing in which a society engages is closely related to its social formation. The symbolic elevation of both alphabetic notation and algebraic numeration come to play central roles in the configuration of modern capitalist society and form an integral pretext to the operations of modern computer technology. This dual mode of phonographic symbolizing in which writing replicates sounds rather than attempting to establish correspondence through an ideographic or pictographic representation allows a limited range of arbitrary signs to produce a system for generating and transmitting meaning.

This mode of writing is institutionalized in the information age through the development of a standard code readable by the machinations of the computer. John von Neuman's contribution to the wave of computerization, in which he suggested that a binary numbering system be used in the building of computers, did little to stop the centralization of an alphabetic-algebraic code of symbolizing. After all, the two digits (or bits) representing "on" and "off" are still represented by the algebraic "1's" and "O's" and in some cases the alphabetic "ones" and "zeros." The computer is integrated into the privileged form of representing by the introduction of the 7- and 8-bit coding of alphabetic characters (26 lower and 26 upper case letters) and the algebraic decimal characters (1–10) as well as a number of other characters representing important functions (=, +, -, x), signifying marks ($, #, @), and punctuation (?, ;, !). These codes are commonly known as the American Standard Code of Information Interchange (ASCII) and in IBM machines, the Extended Binary Coded Decimal Interchange Code (EBCDIC). Translating the alphanumerals into discrete values of binary machine language allows the computer to enter common usage. The computer becomes perfectly suited to carrying on the mode of symbolizing which is most closely correlated to the symbolic demands of the modern political economy and fulfills the requirements for modern governmentality.

This does not mean the computer is limited to the constraints of the dominant form of symbolizing. It engages in a set of exercises which hyperextend these practices into new techniques of power. Through its complicity with the production of meaning and value, the computer is intimately tied up in the processes of credit, eligibility, identification, ownership, spatiality, and temporality, and thus the constitution of modern society and its organization of power. As a form of writing technology, computerization stores its marks over durations of time. Connected to the new techniques of modulating over electrical-based telecommunication circuits, computerization moves its characters rapidly over geographical terrains. It lists any number of objects or people complete with corresponding data fields adding interpretation and description. The computer co-relates. It links, makes connections, and combines people with places and resources.

6. DREAMING ELECTRONIC MONEY

The money sign emerges in modern mass culture as the embodiment of possibility, the semiotic of success. "Money is a dream," said David Bazelon. "It is a piece of paper on which is imprinted in invisible ink the dream of all the things it will buy, all the trinkets and all the power over others" (Bazelon 1963, p. 68). Money no longer needs to have intrinsic value as it did in previous ages. Foucault (1971) historicized the representations of money in the classical age and outlined the variables of signs necessary for representations during that period, or what he was calling at the time an "episteme." He argued that money could only represent wealth during that time, when it itself was precious.

Engaging Goux's essay on "Numismatics," where he recounts a historical commencement of traditional money, where one commodity among others is placed as the unique measure of the values of all other commodities, we see that money becomes a regulator of value. It settles the contradictions of multiple equivalences, governs the exchange of commodities, and settles accounts as a means of payment.

The new circulations of the words information and the information society seems after analysis to institute insidious restrictions of the flows of, if not information, then signification. In short, a discourse or knowledge emerges which delimits or restricts the flow of meanings associated with the standard. It moves towards the creation of a sovereign stock of signs and meanings and maintains it so as to create the units of exchange.

7.   COMPUTER LOGOCENTRICISM AND ORGANIZATIONAL SOVEREIGNTY

This knowledge about money is central to modern capitalism, and integral to it is the codification and technicalization of information. It implicates agents, both people and organizations in its expression. Techniques and language reinforce the new power of money, particularly the new disciplines of accounting and budgeting. This money knowledge is transnational and hegemonic. Corporations and governments both contribute to a process of reproducing money as well as suffer from its structuring effects.

Organizational society has developed specific languages, techniques, and technologies for the legitimation and manipulation of the money sign. A select group of institutions exist in a privileged stratum within the transnational regime of industrial formations and disciplinary practices which allocate power and possibility in modern society. Advancement in the hierarchical scheme of this regime requires the careful coordination of the central sign of contemporary society, the money sign. The money sign

in its various forms such as assets, credit, bonds, stocks, etc., has

become the primary vehicle for organizing modern society.

Technocratic organizations are highly sensitive to the power of

money and the rules through which access to them are constituted. For

the modern corporation, the "bottom-line" is a mega-metaphor.

Corporations strive to display capital accounting results in an organized

and consistent manner to directors, stockholders, and the media. The

word "consistent" here is used because while the corporation is organized

for the surplus accumulation of wealth in the ultimate form of the money-

sign, it is, as Galbraith pointed out long ago, the logic of bureaucratic

capitalism that profits should be rendered predictable (Galbraith 1967).

Money is the organizational lubricant for business and government.

It clearly has taken on new forms and functions in this transformed

technological environment. The practices of budgeting across

multinational divisions, managing treasury accounts, and raising

investment capital have increasingly relied on new forms of computerized

financial representation and substitution. Budgeting becomes an

essential technique of control in the technocratic organization, both in the

corporation and the government. The steeper the organizational chart,

the more budgeting is used as a political tool. Budgeting frames the organization and its objectives in quantitative money terms. It allocates resources to some and restricts them to others. It can be used to redirect programs and to pressure subordinates and organizational adversaries to conform to new objectives.

These money languages provide an understanding of the phenomenon of multimediated information and its importance for the global economy. Combining the powers of computers with the sophistication of a traditional discipline creates a new type of knowledge, one that requires interpretative skills based on a discipline's foundation. The organization of knowledge imposed by various state-of-the art disciplines combined with the new information techniques improves command over the financial, logistical, and productive aspects of the modern organization.[3] With the computer-based technique and knowledge-based interpretive skills becoming intertwined, it behooves to explore some examples of its social impact. The spreadsheet is exemplary; its ability to manipulate numerical data in tabular format and create a variety of calculative scenarios is having an enormous impact on an assortment of disciplines. Since the first spreadsheet was designed in 1978 by a Harvard student to project the consequences of one company's acquisition of another, the application program has become second only to word processing in its popularity. Armed with the new computerized spreadsheets, multiple what-if scenarios can be produced to simulate organizational decisions. The spreadsheet both democratizes numerical power and modeling processes as well as gives numerous advantages to those who can combine its power with other organizational factors such as access to capital. The last decade of spreadsheet capitalism combined the eurocurrency investment capital along with the capabilities of entrepreneur raiders using the spreadsheets to analyze takeover targets and shake up the predominate corporate structure.

The changes in the world economy are complex and the diagnosis as well as prescriptive proposals are contentious. Much of the discussion regarding the changes in the global political economy makes the argument that the Fordist-Keynesian regime of social organization, which combined new productive capacities and new forms of social and macroeconomic regulation, has been in a rapid state of disintegration since the early 1970s. This situates the technocratic information standard, the new computer logocentrism, in the center of the new post-Fordism regime.

8. INFORMATION, MULTIMEDIA, AND THE PRODUCTION PROCESSES

The global economy has shifted mass production to lower-wage and/or automated economies leaving the rest of the world economy to resort to a new type of flexible commerce dispersed among many nations and cultures. The world economy is now undergoing what Harvey (1989) calls a "time-space compression" due to new permissive technologies such as jet airplanes and telecommunications. This has meant a shift from vertically-organized corporations to new networked economies which privilege interorganizational ties by such means as outsourcing and subcontracting. The rigidities of Fordism have been shaken largely by the capabilities of the new information technology–driven organizations. Spatial and temporal dimensions of the economy are being reorganized in the need to reduce turnover times for flexible production and marketing strategies on a global scale. For example, coordination of the logistics of containerization, inventory control, and packaging needed to compete in the new marketplace requires contact with a wide of array of competing services. Access to information such as the GATT agreements on pesticides and food additives as well as other nontariff barriers is becoming critical for devising new global trading strategies.

Most countries have been establishing electronic trading links with other countries in order to tie in a steady flow of orders for its export commodities. This takes a labor and management pool who are practiced with the new techniques of information management. Customer and supplier relationships are increasingly becoming locked into electronic document standards such as EDI (Electronic Data Interchange). International trade which involves customs declarations, cargo clearance, customs duties, delivery orders, insurance certificates, bills of lading, interconnecting carriers, vehicle booking, sales tax, excise tax, quarantines, etc., all consist of document procedures which are becoming increasingly electronic. This has meant a new intensification of mental labor activity and has accelerated the need for new types of labor learning strategies.

One of the most extensive series of studies looking at the new types of multimediated information produced by the convergence of information technology and new work environments was done by Shoshana Zuboff (1988). While her study lacks sufficient linkages to the motivations of the global economy and is phenomenonological-based, thus raising some poignant ontological questions, her insights into the transformations occurring in the workplace are instructive nonetheless. In her book, she argues that the new information technologies have two faces; not only do they automate, but they also informate. That is, what computerization does best, in addition to monitoring automation, is to produce information.

Automation will continue to be an integral part of the industrial equation, but the term "informating" considers the facility of textual and symbolizing processes in the workplace. Zuboff coined the term to refer to the production of symbolically coded information about automated activities. Computers in the modern workplace take three-dimensional tasks and objects and translate them into symbolic data which are presented on a screen. Toil is becoming progressively symbolic and abstract. Labor that was once done manually is now mediated by the computer. Automation, which once decreased the importance of knowledge in the workplace, now becomes heavily dependent on it.

The barriers to entry into the workplace are rising. Physical jobs have become mental jobs as workers move from the floor to the office. The control room has replicated itself from the television studio to the factory office. Workers are tied to the screen as the dominant channel of working activity. Nutured through the television age, the screen is, in a Lacanian-mirror sense, a motherly form to the modern worker. Automated plants producing cars, computers, or paper, do not run themselves, and at the least, require such a high level of investment capital that to sit idly for lack of a machine replacement or a proper diagnostic program is a very expensive oversight. A lot of work is still required to check and maintain equipment, but even these jobs are becoming highly intellective.

The intricacies of the modern logistical capabilities depend on the new symbolic codifications of information. Federal Express, which got its start in the courier business, moved quickly into the package delivery business after the introduction of the facsimile machine dramatically reduced the need to hand deliver important business documents. They now offer package delivery to their clients as well as inventory and parts management systems to facilitate the replacement of sold stock or irreparable equipment. Extending the informating process but also adding a new symbolic coding system allows them an unprecedented ability to offer logistical services. Currently they use Internet and the Word Wide Web to track deliveries.

9. THE UNIVERSAL PRODUCT CODE

One of the most striking and ubiquitous symbolic technologies in use is the Universal Product Code, otherwise known as the "bar code" or as they are known in the industry, "symbologies." This rather inconspicuous set of markings is becoming evident on retail items, distribution packaging, as well as on the literary and public psyche. The generic coding language based on a combination of alphanumeric symbols and machine readable bar codings is fast becoming central to the accelerative logistics and production systems that are now becoming operative in the global political economy. Combined with the interpretative and calculative capabilities of the computer, the bar coding writing system is fast becoming a central means of information coordination.

The bar code is the front end of an elaborate computer information system. It has become an efficient replacement for human-mediated data entry systems through the use of laser scanners which can read the complex codes. Federal Express's Cosmos computerized "supertracker" system in Memphis can get reports as to the exact location of their packages through the bar codes ("Setting the Pace" 1988). Symbologies on each package are scanned at key points in the distribution trajectory. Information from every pickup and delivery is sent to the computer system and can give real-time reports on delivery routes, load factors, the number of packages requiring special handling, etc. They even offer inventory management systems which bypass the wholesaler and deliver products to the customer in twelve to seventy-two hours. Instead of maintaining warehouses of inventory, these systems allow just-in-time deliveries to industrial or retail locations. Bar coding has become an essential part of the accelerative logistics and production systems that are now becoming operative around the world.

The technology for storing data in this form is developing rapidly. A variety of formats have been produced which, instead of competing with one another, are being allocated for a variety of applications. One corporation, Symbol Technologies, has created a new bar code called PDF 417 (Portable Data File) which can encode nearly 2,000 alphanumeric or 3,000 numeric characters in just a few square inches (*The Bookseller* 1991, p. 1583). The three kilobytes has been touted as containing enough memory to store the Gettysburg Address and takes less than a second to scan. Some of its applications include medical histories, manufacturing instructions, parts tracking, as well as equipment calibration and maintenance encoded right into the machines.

10. INFORMATION STANDARDS AND SOVEREIGNTIES

We can now return to Wriston's thesis that an information standard has emerged to dominate the international money system to the extent that nation-states and corporate bureaucracies are seriously threatened and are being replaced by a new form of electronic economic and political democracy. In this sense, the information standard as presented by Wriston is a contribution to the understanding of how technology has been integrated into the international political economy, especially since he alludes to its more generalized form and its translation and substitution into other symbolic forms.

Adam Smith's effectiveness was in part due to his ability to articulate a model of the individual which, while not necessarily novel for his time, nevertheless, resonated with the changes from an absolute state to a government based on legislative, judicial, and executive powers. Unlike Thomas Hobbes, whose individuals surrendered knowingly their will to the sovereign in order to obtain civil peace, the individual in a Smithian universe is a socialized being with an "attenuated" sovereignty, "not a self-contained, sovereign actor but a bifurcated or double self, containing both an actor and an imagined observer through whom action predicates are mediated" (*The Bookseller* 1991, p. 453).

What this means is that the static economy organized around mercantilism "becomes unstuck." Sovereignty and stasis give way to circulation and mobility as exchange is favored over the maintenance of a monarchical realm. A new form of sovereignty, that of private ownership and accumulation, emerges to become the new "wealth of nations." Sovereignty is subjugated to the new flows and circulations of the economy. Likewise, the ideal spiritual heirarchy is also subsumed in the energies of everyday life (Shapiro 1990, p. 2).

Shapiro is quick to qualify his use of the European Monetary Union as an example for examining the sovereignty-exchange nexus less the reader come to the conclusion that these impelling forces continually work against each other.

On an initial inspection, the worldwide phenomenon of privatization may appear to champion the forces of exchange against a history of bureaucratic control over the economy. Indeed, the rationalizations which enter public discourse about selling government facilities are often riddled with references to free enterprise, unrestricted communications, and the dispersing of wealth. However, it is unclear and unlikely that the forces of sovereignty would lie down so sheepishly. The technological imperative is fraught with other forms of control and collectivities. As this chapter strove to point out, the emergence of the new computerized information standard is a new form of signifying which must necessarily preclude other forms of signification. A new stabilization of sovereignty and its resultant solidification of meaning preempts other forms of otherness and consequently what is also at stake are the new identities and "proto-sovereignties." The twilight of national sovereignty suggests that this stable form, with all its administrative posturing and violence, may soon give way to other forms of political and economic authority. It is not too early to raise substantive questions about the new domains and subjectivities which will rush in to take its place.

Editor's Note: In an age of information superhighways, cyberspace creates many complications which the twenty-first century will have to face. One example is the emergence of a National Computer Crimes Squad established by the FBI in the United States. Computer security experts foresee a growing need for law enforcement along the electronic superhighway. As the "Information Standard," discussed in the chapter, becomes more universal, more security concerns regarding electronic funds and other forms of strategic information exchange will gain prominence in cyberspace.

NOTES

1. For an excellent summary of the events leading to the "Third World Debt Crisis," refer to Moffit (1983) and Pennings (1986).

2. Cees Hamelink, Herb Schiller, V. Mosco, A. Gouldner, and J. Glatung, are excellent writers from this perspective. Although not Marxist-informed, Majid Tehranian's, *Technologies of Power: Information Machines and Democratic Prospects* is one of the most potent critical texts in this field.

3. For an extended argument, see my commentary in *Media Asia*, no. 4, 1988.

REFERENCES

Bazelon, D., 1963, *The Paper Economy*, Random House, New York.

The Bookseller, 1991, November 29.

Setting the Pace for the 1990s, Federal Express Corporate Film, Greg Cooke Productions, Inc., 1988.

Dordick, Herbert S., 1986, *Understanding Modern Telecommunications*, McGraw-Hill, New York.

Dordick, H. S., and D. Neubauer, 1985, "Information as Currency: Organizational Restructuring under the Impact of the Information Revolution", *Keio Review*, no. 25.

Finlay, M., 1987, *Powermatics: A Discursive Critique of New Communications Technologies*, Routledge and Kegan Paul, London.

Foucault, Michael, 1971, *The Order of Things: An Archaeology of the Human Sciences*, Pantheon Books, New York.

Galbraith, John K., 1967, *The New Industrial State*, The New American Library, New York.

Goux, Jean-Joseph, 1990, *Symbolic Economies*, translated by Jeniffer Gage, Cornell University, Cornell Press.

Harvey, D., 1989, *The Condition of Post-Modernity*, Basil Blackwell, Oxford.

Ito, Youichi, 1981, "The 'Johoka Shakai' Approach to the Study of Communications in Japan", in G. C. Wilhoit and H. De Brock (eds.), *Mass Communications Review Yearbook*, vol. 2., Sage, Beverly Hills, CA.

Masuda, Yoneji, 1981, *The Information Society: A Post-Industrial Society*, Institute for the Information Society, Tokyo.

Moffit, M., 1983, *The World's Money: International Banking from Bretton Woods to the Brink of Insolvency*, Simon and Schuster, New York.

Pennings, Anthony, 1986, *Deregulation and the Telecommunications Structure of Transnationally Integrated Financial Industries*, MA Thesis, University of Hawaii.

Porat, Marc U., 1977, *The Information Economy: Definition and Measurement*, vol. 1., OT Special Publication 77–12(1), Department of Commerce, Washington, D.C.

Shannon, Claude E., and Warren Weaver, 1949, *The Mathematical Theory of Communication*, University of Illinois Press, Urbana.

Shapiro, M., 1985, Sovereignty and Exchange in the Orders of Modernity, *Alternatives*, vol. 16, pp. 447–77.

——, 1990, Moral Sentiments and Textual Subjectivities: Smith with Sade and Postmodernism, paper delivered at the 86th Annual Meeting of the American Political Science Association, August 30–September 2, 1990, San Francisco, California.

Tehranian, Majid, 1990, *Technologies of Power: Information Machines and Democratic Prospects*, Ablex Publishing Company, New Jersey.

Weiner, N., 1954, *The Human Use of Human Beings: Cybernetics and Society*, Anchor, New York.

Wriston, W., 1992, *The Twilight of Sovereignty: How the Information Revolution is Changing our World*, Charles Scribner's Sons, New York.

Zuboff, Shoshana, 1988, *In the Age of the Smart Machine: The Future of Work and Power*, Basic Books, New York.

GLOBAL TELECOMMUNICATIONS STANDARDIZATION IN TRANSITION*

DAVID LASSNER

Global
L96
L98
F02

1.   ABSTRACT

Global telecommunications standards are a crucial aspect of the increasing internationalization and informatization of the world today. They provide important benefits for users and providers alike. International telecommunications standards have traditionally been established in international forums. While all nations are eligible to participate in standardization activities, the outcomes are determined primarily by the industrialized world with dominant roles played by multinational manufacturing interests and telecommunications operators. Involvement has been minimal by lesser developed nations and others.

Recently there has been an increasing level of telecommunications standardization activity by regional/national organizations and industrial alliances, with these programs centered in Japan, the United States, and Europe. This movement of activity to forums outside the formal global arena has the potential to institutionalize the limited involvement of the nonmanufacturing nations. At the same time, these new telecommunications standards are crucial in providing advanced services that help support economic and social development through the provision of a modern public telecommunications infrastructure.

As the world faces the development and maturation of a wide range of new, public high-speed and low-speed digital telecommunications technologies, what is the impact of the lack of involvement of nonparticipating nations in standardization? If there are negative impacts on these nations or the standards themselves, what can be done to remedy this nonparticipation as the new standardization regime emerges? These questions are examined particularly in the context of three nations in Southeast Asia, namely, Indonesia, Malaysia, and Singapore.

2.   INTRODUCTION

Telecommunications standards specify the interconnections among different system components. In effect, these are the rules of developing and deploying interoperable devices and networks. One of the great successes of telecommunications standardization is the international wired telephone system. Standards specify how telephones connect to phone switches, how phone switches are interconnected, and how all these devices interoperate, even across borders, to provide transparent end-to-end voice communication service to users. Similarly, facsimile machines communicate in standardized ways over this network to provide global image transmission capability. This level of international agreement has not been achieved with video; multiple color television standards prevail and prospects for a single standard for high definition TV (HDTV) are converging only after multiple starts and fits in all regions of the world. In high-speed data communication and sophisticated digital information networking applications, a variety of competing and conflicting interconnection technologies abound. And a variety of standards for mobile telephony constitute a formidable barrier to global roaming services based on currently deployed technologies.

The lack of transparent connectivity with the latest technologies threatens to be problematic in the face of emerging economic and technological trends. Communications and information technologies are increasingly the basis of economic activities as well as being an important economic sector themselves. With the trend toward digitization of communications and rapid advances in digital hardware and software technologies, there is a convergence between the worlds of computing and communications that is rapidly taking center stage in both industries. Economic activity is increasingly taking place on a global rather than local or national basis. The names of major regional and global trade agreements, such as GATT and NAFTA, have even entered the popular vernacular. And in the telecommunications sector, there is a worldwide increase in privatization and liberalization of previously monopolistic government service providers.

All this adds up to a greater need for telecommunications standards than ever before in order to support more international communications in an increasingly complex competitive environment.

3.   THE ROLE OF STANDARDIZATION

Increasing standardization is almost universally considered to be a positive trend. Users of information and telecommunications systems are major beneficiaries of open standards. Following are some of the key benefits, which are described in the economic literature on standards (David 1990, Farrell and Saloner 1986, Besen and Saloner 1989, Besen and Johnson 1986).

3.1. Interoperability

Interoperability with large numbers of other compatibly equipped users provides significant positive externalities; i.e., telecommunications networks become more valuable with an increase in the number of users that are reachable through them. For example, users of telephones, facsimile machines, and dial-up modems find greater use and value in these services since their investment in equipment typically allows them to reach other users of the same service.

3.2. Interconnectivity

Standardized interfaces and services make it much easier and less expensive for users to design and construct complex networks using a mix of equipment and telecommunications services.

3.3. Decreased Costs

Standardization permits economies of scale in production, for example, by making mass production of specialized chip sets economical. Furthermore, the presence of multiple vendors of standard products in the marketplace leads to more competitive pricing and lower costs than in single-source markets.

3.4. Portability, Scalability, and Vendor Independence

Standardization makes it possible for users to move applications and services to new equipment and providers without major costs. As requirements grow, capabilities may be increased gradually and gracefully with less reliance on a particular equipment or provider. Users perceive value in the increased flexibility and decreased risk associated with portable applications; they need not be dependent on the fortunes and practices of a single vendor.

While a successful proprietary standard that permits monopolistic dominance of a market is every vendor's dream, it is apparent that this is no longer possible in the modern global telecommunications environment. Without such ownership of a dominant proprietary standard, most vendors see great benefits in open standards. Acceptance of this reality is evidenced by active participation in a wide range of standardization forums by nearly every major telecommunications manufacturer and provider. Benefits to vendors of standardized markets over technologically fragmented markets include:

1. Expanded markets. Adherence to standards can increase the scope of a vendor's market. In markets without standards, a vendor may only be able to sell to customers who have selected their particular version of a technology. However, all users of a standardized system are potential customers of a vendor with products and services that adhere to those standards. This is particularly important in international markets, in which large vendors perceive adherence to global standards as key to becoming viable suppliers in other countries.

2. Simpler network design and connectivity. Telecommunications vendors are often called upon to design and build customer networks. It can be more difficult to guarantee successful technology deployment and thereby close a sale with a customer without standards to govern how to interconnect new equipment and services with older existing networks.

3. Reduced R&D costs. Use of standards that are developed collaboratively and shared openly can dramatically reduce research and development costs for individual vendors in the marketplace. Standardization provides manufacturers with access to technologies and ideas from sources outside their own labs (Onufrak 1992).

Both users and vendors also benefit from the manner in which standards create a stable technology infrastructure for future research and development. For example, the emergence of the IBM PC as a de facto standard for personal computing enabled a wide range of after-market products that rely on the standard. This spurred both hardware and software markets, creating business opportunities for vendors and improved services for users. Conversely, the lack of standards is perceived by vendors and service providers as a major barrier to increased markets for new services and technologies. Users are cautious about investing in nonstandard telecommunications technologies.

Standardization of the public network is especially important. In the United States and other countries with separate and substantial cable television and telephone industries, standardization will be required for either type of company to be able to provide the full range of integrated communication and information services (voice, data, and video) that will provide markets adequate to cost-justify the deployment of fiber to the home.

Standards for advanced high-performance services are required if telecommunications providers will be able to meet the needs of businesses and serve them with public networks rather than watching the continuing proliferation of private networks. And in less developed countries, implementation of standards for advanced data services on the public network will be required to prevent bypass by businesses using VSAT or other off-network technologies. Business customers are absolutely needed if public telecommunications operators (PTOs) in these countries are to have the revenue base necessary to fund the infrastructure development they require.

Standardization is, however, sometimes seen as a barrier to technological advancement. Adoption of standards may make it difficult or impossible to migrate to superior technologies as they become available. Use of the QWERTY keyboard layout, even in the face of superior alternatives such as the Dvorak keyboard, is an excellent example of this inertia effect. Standards nearly always lag behind the most advanced technologies. Non-adherence to standards frees vendors to take advantage of the latest developments in its product line, offering more advanced applications and products to users willing to purchase them before the establishment of standards in new areas.

4.   TRADITIONAL STANDARDIZATION

Global standards were traditionally set by formalized international standards organizations. These standards often described the basic interconnectivity of systems, leaving some details of user connectivity to be determined on a national or regional basis. An important common thread in telecommunications standardization has been the dominant role of the PTOs and their preferred equipment suppliers. Usually the PTO was a government entity (in most countries a PTT) or in some cases, a regulated monopoly (e.g., AT&T in the United States). The preferred equipment supplier has likely been a national monopoly or the PTO itself. The most important international standards organizations have been the International Telecommunications Union and the International Organization for Standardization.

The International Telecommunications Union (ITU) is a treaty organization that specializes in telecommunications (ITU 1992). Although it now operates under the auspices of the United Nations, the beginnings of the ITU go back to the mid-1800s and the development of the international telegraph network. Full formal membership is restricted to governments, and most nations of the world retain membership. Until 1993, standardization activities took place in two permanent bodies within the ITU: the International Telegraph and Telephone Consultative Committee (CCITT) and the International Radio Consultative Committee (CCIR). The CCIR was primarily

responsible for transmissions that occur through radio transmission, while
CCITT was responsible for wired telecommunications standards. In
recognition of the increasing overlap between these two areas, particularly
with the increasing use of digital transmission systems, the ITU has now
consolidated its standardization work in a single Telecommunications
Standardization Bureau (ITU 1993). Most of the actual standards work in the
ITU is done in working groups made up primarily of private sector entities
that are identified by their governments as recognized private operating
agencies (RPOAs) or scientific and industrial organizations (SIOs).

The International Organization for Standardization (ISO) is an
international voluntary body that promotes and facilitates standardization
(ISO 1991). Membership is open to the national standards organization of
each country, which may or may not be a government entity. A separate
body, the International Electrotechnical Commission (IEC), is responsible for
electrical and electronic standardization; all other areas fall into the purview
of ISO. Together, ISO and IEC have formed the Joint Technical Committee 1
(JTC1) with overall responsibility for their information technology
standardization program.

ISO and the ITU are headquartered across the street from each other in
Geneva. Typically telecommunications standardization has been left to the
ITU, and information system standardization has been handled by ISO.

There is, of course, significant overlap between these two areas and the two groups work closely together to avoid unnecessary duplication of effort. For example, in the crucial area of open systems interconnect (OSI), many of the key standards documents issued by both bodies are identical.

Standards-making processes vary according to the organization developing the standards, the nature of the standard itself, and the state of development of the particular technology in question. A typical process begins with the decision, generally made at the highest level of the standards organization, that a standard is required in a particular area. The task is then assigned to a particular study group or technical committee, usually composed of technical staff from members with an interest in that technology. The technical group works until it reaches consensus. Then the standard is ratified in some fashion back at the highest level of the overall organization. The final product of the process is generally considered to be a recommendation. It only has force of law if a legal authority within a particular country or region mandates use of the recommendation as a required standard.

Sometimes consensus simply cannot be reached. This is usually due to an inability to agree on a particular standard for the technology in question. Even when there are overriding and pervasive benefits to standardization, this can occur if the participants have strong vested interests in a particular standard or in blocking a particular standard. When, however, a standard is finally recommended, it may be the result of one of the following kinds of processes:

1. Standards construction. Standards may be developed from the ground up. This can be a slow, negotiated process, and many compromises may take place along the way. This takes place most often when standardization begins in advance of actual development and deployment, before participants in the standards process have a strong vested interest in promoting or blocking a particular outcome, e.g., their own or a competitor's product respectively. Standardization of Integrated Services Digital Networks (ISDN) is an example of this kind of activity. There are several potential dangers with this kind of early standardization. One is that by standardizing before a product or service is proven in the marketplace, it may turn out that there is little demand by users. Another problem is that standards set in advance of implementation may be technically flawed due to the lack of real-world experience with the technology and may require significant revision or enhancement.

2. Standards adoption. In some cases an existing implementation is accepted as-is as a standard. This may be a quick way of establishing a standard, but it may give an advantage to the developer whose standard is selected. Competition among various

implementations to be recognized as the standard may be fierce, and participants in the standardization process who have no favored implementation may simply wish to block someone else's proposal to avoid yielding a competitive advantage.

3. Standards adaptation. A new standard may be the result of a process somewhere between the previous two. An existing implementation may be the basis of a negotiated standard, after being adapted in accordance with the demands of other participants. The changes may be required to improve the standard functionally or simply to limit the competitive advantage of the original proposer.

4. Standards with options. Sometimes there can be no adaptation or compromise that garners consensus. In that case, some organizations will accept standards with multiple options, where each option corresponds to one of the proposals. This outcome is slightly more effective than no standardization at all, since it may limit the number of options in the marketplace and permit some of the benefits of standardization.

The traditional process of developing international standards has been widely criticized. Some of the most common and serious issues raised regarding the process itself are:

1. Slowness and bureaucracy. The formal standards setting process

   has been considered to be slow and burdened by bureaucracy.

   Working methods need improvement, perhaps using

   telecommunications technologies (Rutkowski 1991). Even with new

   "fast track" procedures in effect, beginning a new standardization

   process after the need is perceived or competing technologies are

   already on the market is too late (CCIT 1989). With the rapidly

   accelerating pace of technological advancement, this problem has

   become even more severe. In a competitive marketplace, vendors

   will simply not wait for standards before developing and deploying

   new technologies that users need. It is now widely recognized that

   standardization must be more market-oriented and efficient (Irmer

   1990).

2. Inability to reach a consensus. Often participants in the

   standardization process will arrive with agendas that make it

   impossible to reach consensus. This might be due to prior

   investments in particular implementations of a technology, in which

   case multiple participants may have vested interest in seeing their

   implementations selected as the standard. Or one or more

   participants might simply wish to block some other participant's

   proposal in order to prevent that company or country from obtaining

   a competitive advantage through their technological headstart.

Videotex and color television standardization efforts are examples of the problems that occur when entrenched national positions are irreconcilable in international forums (Savage 1989).

3. Quality of negotiated standards. As standards are negotiated, particularly in the political setting of an international forum, technical excellence may be sacrificed to the pragmatic need for agreement and political considerations unrelated to the standard or technology under study (Rose 1990). Standards developed in advance of implementations, while perhaps making it easier to reach consensus, may be technically flawed and require significant revision as lessons are learned through implementation efforts.

4. Resource requirements and distribution of costs. Active participation in the standards process is extremely expensive. It is estimated that hundreds of millions of dollars per year are spent on standardization (Tarjanne 1990). Lengthy meetings are held in international locations. In addition to the significant travel and lodging costs that must be borne by the participants' employers, there are the costs associated with the time and salary of these highly valued employees. Some active participants in the standards processes believe that the costs of standards development are unfairly allocated; that while they bear the bulk of the costs,

numerous other nonparticipants have equal rights to develop products in accord with the resultant standards (Berg and Schumny 1990). These nonparticipants benefit from the research and expense of the standards process without paying any of the costs.

5. Limited access to standards. The language of standards is sometimes thought to be obtuse; new terminology is often devised, critics claim unnecessarily. The standards were traditionally only available in hard-copy, generally at prices high enough to serve as a deterrent to casual collection and inspection (Savage 1989). The ISO clearly states that their publication resale program helps support the organization. The ITU now makes its standards available on-line and on CD-ROM (ITU 1993).

6. Intellectual property rights. All cooperative standards organizations grapple with the relation of standards to intellectual property rights (Irmer 1992). Standards are expected to be freely available and widely used, but there is the possibility that compliance with a particular standard may require use of a technology (e.g., a patent) that is owned by an individual or a firm. This would clearly compromise the standard and potentially the standardization process. Most standards organizations ask that participants openly declare their interests in any intellectual property that might be

required for compliance with standards under consideration.
Participants with such interests must be willing to agree to license
their interests on a reasonable and nondiscriminatory basis or work
on the standard will not proceed. However there still remains the
possibility of unintentional nondeclaration, malicious
nondeclaration, or overlap between a standard and some intellectual
property right of a nonparticipant in the standardization process.

7. Limited participation. Users seldom have the resources to be active
   players in the standards forums, although organizations such as the
   International Telecommunications Users Group (INTUG) and the
   International Chamber of Commerce (ICC) can serve as a voice for
   users. Similarly, lesser-developed nations also lack the resources to
   actively participate and it is sometimes claimed that the scope
   and/or contents of the standards being developed do not address
   their needs. Standards bodies are attempting to bring users into the
   process but are still grappling with how to best do so (Knight 1992).

5.   CHANGES IN STANDARDIZATION

The quickening pace of technological advancement, increasing
internationalization of business, and worldwide trends toward deregulation
and liberalization of telecommunications have relentlessly spurred the

demand for more open and timely global telecommunications standards. The ITU undertook a major strategic planning process in order to identify the basis and direction for needed structural reform (ITU 1991). This restructuring and both preceding and proceeding reforms in the standardization process have helped the ITU respond to the changing environment (Irmer 1994). Similarly, the ISO and IEC have reconsidered their standardization processes in the face of fundamental changes in the technological environment (ISO/IEC 1990).

One of the most important trends in standardization through the mid- and late 1980s has been the emergence of what are generally referred to as regional standards organizations (RSOs). These organizations have grown out of the recognition that with the deregulation and liberalization of telecommunication authority, formal national or regional standards are required to ensure interoperability of public telecommunications systems. International telecommunications standards generally do not address all aspects of the system (e.g., the user-network interface) and sometimes leave options from which each country can choose. Therefore some standardization is still required on a more localized basis.

After the divestiture of AT&T in the United States, it was clear that there would need to be coordination among the new carriers to assure continuing network integration and interoperability (Lifchus 1985). The

Exchange Carriers Standards Association (ECSA) was created to meet this
need. With the recognition that carriers alone could not set telecommunications standards, a more broad-based standards committee was formed
under the auspices of ECSA. Beginning in early 1984, Committee T1
assumed responsibility for telecommunications standards in the United
States. This committee, which is made up of representatives of the carriers
themselves as well as manufacturers, interexchange carriers, vendors, users,
and others with an interest in telecommunications standardization, is
responsible for wired telecommunications standards.

Japan began deregulation of telecommunications in 1985. As in the
United States, it was recognized that standardization should be primarily a
nongovernmental activity. The Telecommunications Technology Committee
(TTC), similar to T1 in the United States, was established in late 1985
(Habara 1990).

The European Conference of Postal and Telecommunications
Administrations (CEPT) was traditionally responsible for European
telecommunications standards. As part of the impetus to create a single
European market by 1992, the issues surrounding telecommunications were
carefully studied. A fully integrated telecommunications system was
considered crucial both for the role of telecommunications in facilitating the
single market as well as due to the size and importance of the

telecommunications industry itself (equipment and services). Regional standards to supplement international standards were a clear requirement, so in 1988 the new European Telecommunications Standards Institute (ETSI) was created with responsibility for both radio and wired telecommunications standardization (Besen 1990). Unlike T1 and TTC, ETSI includes administrations (governments) and national standards bodies in its membership. It also engages professionals on a for-hire basis to engage in standards development, as opposed to relying on voluntary contributions from the staff of its members.

RSOs were not established to supplant the global standardization efforts of organizations like the ITU. Rather their intent was to set standards in areas in which there are no international standards, and establish national or regional profiles in areas where the international standards are incomplete or allow options. All the RSOs recognized the importance of global standards, both to assure interoperability of worldwide networks and to facilitate access to worldwide markets by their own telecommunications sectors. However, as they began operation, it became clear that there was tremendous potential for conflict among the RSOs and ITU, or at the very least, for expensive and unnecessary duplication of effort.

Seeing the need for coordination, these three RSOs and the ITU met together at the first Interregional Telecommunications Standards Conference in February 1990 in Fredricksburg, Virginia. Generally avoiding the political

issues of formal relationships among governmental and nongovernmental as

well as regional and national organizations, the meeting, which has been

dubbed the first "standards summit," was considered a success. A second

interregional meeting was held in September 1991 in Nice and a third in

Tokyo in 1992, where the group evolved into the Global Standards

Collaboration (GSC) (Habara 1994). The overall outcomes of these summits

are: (1) reaffirmation of the preeminence of the ITU in global telecommu-

nications standardization, (2) agreement to continue working together to

improve the efficiency and effectiveness of overall standardization efforts,

(3) agreement to use electronic communications technologies to improve

interchange among the groups, (4) agreement to work toward a coordinated

work program and schedule, and (5) agreement on a number of high-priority

areas in need of standardization.

The ITU agreed to not begin developing standards in areas under active

study by the RSOs and the RSOs agreed to work with each other to attempt

to avoid situations in which conflicting and competing proposals are

presented to the ITU. Improved interaction through use of information

technologies coupled with a coordinated work and meeting schedule is

expected to improve the efficiency and timeliness of the standardization

process. Still, the ultimate impact of the RSOs and standards summitry on the role of the ITU remains unknown (Besen and Farrell 1991, Hawkins 1992).

A new trend toward the creation of industry alliances for standardization should also be noted as an important aspect of the landscape. Corporations are increasingly impatient and unwilling to wait for bureaucracies to function and they recognize the importance of standards if new technologies are to succeed in the marketplace. Very mobile task forces of corporations interested in a particular technology can be formed quickly for specific purposes, to allocate resources, and get on with the work while larger and less wieldy organizations are still discussing needs and processes. Examples of such groupings are the ATM Forum and the Frame Relay Forum. These industry alliances are often international in nature. It remains to be seen how they will fit in with the new regime.

Finally, the Internet software standards development process should be identified as the most dynamic model for standards-making. In contrast to traditional standardization models, Internet standards are developed through individual direct participation in a bottom-up process that actually uses networks to develop network standards (Crocker 1993).

The changes in the standardization landscape over the past decade are profound. There is a faster, broader, and deeper pace of standardization activity, and it is taking place in a growing number of settings. There is now explicit recognition of the work done outside formally constituted international working groups (Rutkowski 1994).

6.   INVOLVEMENT OF LDCS AND NICS

Active participation in the global standards-making activities is quite expensive. Participants must attend numerous meetings away from home. And in addition to the costs of travel, housing, and research, the employers or sponsors of the participants lose their time and the benefits of their skills which could be applied to other projects. The result is that a relatively small number of ITU member nations are active participants in standardization. The lesser-developed countries (LDCs) and the newly industrialized countries (NICs), such as those of Southeast Asia, tend not to have the highly technical human resources required to support highly active involvement in the global standardization activities. And if they do, these are often the people their nations can least afford to do without as they attempt to build their national infrastructures. The new standards summitry has made the long-acknowledged dominance of Europe, Japan, and the United States more explicit. But the standards proposed by any or all of the RSOs arrive at the ITU as proposals and must still be approved there if they are to achieve status as formal recommendations. It can be argued that this provides nonpartic-ipants a point of access into the process that might otherwise be unavailable altogether. Similarly, the industrial forums argue that they are not really standardization bodies but cooperative implementation groups that iron out details omitted from international standards.

One can question whether or not widespread participation in actual standardization activities is important to smaller countries that do not have significant telecommunications manufacturing or export sectors of their own. Just as users believe they need greater input into the process, there is no reason to believe that LDCs and NICs do not share the same sentiment. While, on the one hand, some of the grueling technical work of standards-making is really an engineering function, on the other hand, decisions on priorities and interfaces may have direct impact on services, policies, and infrastructure development. For example, it has been claimed that the developing countries have not been well-served by decisions on modulation techniques for standardized data modems that operate over the public switched telephone network (Baran 1991). Others have questioned whether the overall direction of modern digital technologies, including the Advanced Intelligent Network, will benefit any but the more-developed countries (Schiller and Fregoso 1991).

These issues are more specific instantiations of the general question of whether nonparticipation in standardization by developing countries is leading to inappropriate or inadequate standards to meet their needs. Similar questions include the impact of lost opportunities for human resource development and political or emotional effects of nonparticipation. When faced with competing technology options from different regions of the world, e.g., regarding mobile telephony, nonparticipants need some basis on which to make decisions. Conceivably, these decisions could be made on such factors as protection of (possibly fragile) local telecommunications industries, which multinational corporations have subsidiaries in their country, existing trading partner relationships, and encouragement or discouragement of information transfers across certain borders.

7.   PERSPECTIVES OF NONPARTICIPANTS

Based on interviews with officials responsible for standardization in these countries, it was found that their concerns are much more straightforward than one might expect. The major concern expressed was the requirement for single, effective international standards. The ITU has clearly been viewed as the preeminent authority in this area. In many cases, technology that does not comply with ITU recommendations will simply not be deployed. Although the ITU is concerned with both urban and rural telecommunications, the decisions on what to standardize are made at the Study Groups where the less-developed countries are not well represented due to the resource requirements associated with participation (i.e., money, time, and qualified people). Standards developed are mainly aimed at the needs of the industrialized countries, but these standards are also applicable to the urban needs of the developing countries.

Multiple standards are particularly costly to these countries since they often must deploy multistandard technology and/or translators. In addition, the cost of personnel and expertise to maintain multiple systems is a drain on an already scarce resource. However they find themselves largely powerless to mediate between the major players (i.e., the United States, Japan, and Europe) to resolve these problems in international forums. There has been a certain amount of anxiety regarding the RSOs (T1, TTC, ETSI), but there is also optimism now that they are meeting more often, often with the ITU. The goal of the standards summitry to achieve global standards, not regional standards, is seen as beneficial to both developing and developed nations.

More fully specified global standards would allow easier network design and development. This might also permit developing countries to do more of their own manufacturing. This would not limit innovation and product differentiation since the private sector could still compete with customized systems that have advanced and unique features. These interests must be balanced.

In selecting technologies there is relatively limited interest in the origins of a particular standard. Some countries have tended to align themselves

with technologies from one region or another, but are open to any
ITU-standardized technology. When they do have to choose among U.S.,
European, and Japanese standards, they are less likely to choose a Japanese
technology. This is because the Japanese are seen as more likely to be willing
and able to manufacture to any standard anyway, so a nonmanufacturing
nation broadens its supplier options by choosing a U.S. or European
standard. Also, the Japanese are not perceived as being as strong in
influencing global standardization as is the United States and Europe.

There is relatively limited concern about nonparticipation in
standardization activities, even in countries which are quite active in other
ITU and international activities. These nations attend a limited number of
standardization study group meetings, mostly to stay informed of new
technologies and services. Their major interest is that there be a single,
well-specified global standards rather than the details and specifics of the
standards themselves. Without major manufacturing interests of their own,
they see little need to do much more at this time.

The ITU standards documents are considered difficult to understand for
people who are not actively involved in design and manufacturing, and even
engineers find them fairly opaque. More explanations and improved linkages
between standards and services would be very useful to nonparticipants of

the process so that they can better understand the implications of each standard.

There are also concerns about the frequency management implications of standardization. Frequency management and standardization must be more closely coordinated as new technologies are developed and deployed. Differing regional standards with differing frequency requirements or expectations, such as in mobile telephony, are particularly problematic.

Although full-fledged RSOs may not be appropriate at this time, there do seem to be opportunities for regional cooperation around telecommunications standardization. One role could be as an advocate for improvements in standards (e.g., more specificity, better explanations) that would support the region. Regional cooperation could also improve overall education and understanding of new telecommunications technologies being standardized. And some regional harmonization would be useful, particularly in wireless technologies where spectrum management is at issue and for any mobile communications technologies in which cross-border interoperability is mutually agreed to be desirable.

8.   CONCLUSION

Today's global standardization activities are being driven by the industrialized nations' needs for the timely development of highly technical standards for the interoperability of complex integrated digital systems, the growing influence and needs of transnational interests, and the explicit recognition of the increasing importance of work done by regional and national standards organizations in the developed countries. In the face of rapidly advancing technological development and an increasingly deregulated and liberalized environment, a great deal of important and influential standardization activity lies ahead. For much of the developed world, narrowband ISDN technology is now available, integrating voice and low-speed data. Broadband ISDN is just around the corner, promising to integrate high-speed data and video into the network. Development of advanced intelligent networks is starting, using ISDN as a building block. Open network regimes are coming closer in the largest markets, which will open whole new areas to competition, and wireless personal communications networks are one of the fastest new growth areas in telecommunications.

Nonmanufacturing countries are not playing a major role in the development of the telecommunications standards which will make all this possible. However, their concerns about their nonparticipation are less severe than one might guess. Regionalism in the developed world is already changing the telecommunications standardization regime. Regional activity by nonparticipants may help these nations address their concerns and benefit more fully from rapid advances in telecommunications technology.

NOTE

*   This research has been supported by the East-West Center and the University of Hawaii.

REFERENCES

Baran, Paul, 1991, "The Developing Countries and Data Modems Standardization", *Intermedia*, vol. 19, no. 6 (November/December), pp. 12–18.

Berg, John L., and Harald Schumny, 1990, *An Analysis of the Information Technology Standardization Process*, North-Holland, Amsterdam.

Besen, Stanley M., 1990, "The European Telecommunications Standards Institute: A Preliminary Analysis", *Telecommunications Policy* (December), pp. 521–30.

Besen, Stanley M., and Joseph Farrell, 1991, "The Role of the ITU in Standardization: Pre-eminence, Impotence or Rubber Stamp?" *Telecommunications Policy*, vol. 15, no. 4 (August), pp. 311–21.

Besen, Stanley M., and Leland L. Johnson, 1986, *Compatibility Standards, Competition, and Innovation in the Broadcasting Industry*, R-3453-NSF, Rand Corporation, Santa Monica, CA.

Besen, Stanley M., and Garth Saloner, 1989, "The Economics of Telecommunications Standards", in Robert W. Crandall and Kenneth Flamm (eds.), *Changing the Rules: Technological Change, International Competition and Regulation in Communications*, Brookings Institution, Washington, D.C.

CCITT, 1989, *CCITT IXth Plenary Assembly (Blue Book), Vol. I - Fasc.*, Resolution no. 2, 1.2., CCITT, Geneva.

Crocker, D., 1993, "Making Standards the IETF Way", *StandardView*, vol. 1, no. 1.

David, Paul A., 1990, "Some New Standards for the Economics of Standardization in the Information Age", in Partha Dasgupta and Paul Stoneman (eds.), *Economic Policy and Technological Performance*, Cambridge University Press, Cambridge.

Farrell, Joseph, and Garth Saloner, 1986, *Economic Issues in Standardization*, Sloan WP# 1795-86, MIT.

Habara, Kohei, 1990, Standardization Activities at TTC: Heading Toward New Changes, paper presented at the Symposium on Telecommunications Standards, Tokyo, July.

———, 1994, "Cooperation in Standardization", *IEEE Communications Magazine*, vol. 32, no. 1 (January), pp. 78–84.

Hawkins, Richard W., 1992, "The Doctrine of Regionalism: A New Dimension for International Standardization in Telecommunications", *Telecommunications Policy*, vol. 15, no. 4 (May/June), pp. 339–53.

International Organization for Standardization (ISO), 1990, A *Vision for the Future: Standards Needs for Emerging Technologies*, ISO/IEC, Geneva.

———, 1991, *ISO Memento*, ISO, Geneva.

International Telecommunication Union (ITU), 1991, *Tomorrow's ITU: The Challenges of Change*, Report of the High Level Committee to Review the Structure and Functioning of the International Telecommunications Union, ITU, Geneva.

———, 1992, *The International Telecommunication Union: An Overview*, October, ITU, Geneva,

———, 1993, *TELEDEC Auto-Answering Mailbox: CCITT User's Guide*, January, ITU, Geneva.

182

————, 1993, *The New ITU: Round-up*, January, ITU, Geneva.

Irmer, Theodor, 1990, Telecommunications Standardization at a Crossroads:
The Challenge of the 90 for CCITT, paper presented at the Symposium on
Telecommunications Standards, Tokyo, July.

————, 1992, The Dynamics of Standardization, paper presented at
CommunicAsia 92, Singapore, June.

————, 1994, "Shaping Future Telecommunications: The Challenge of Global
Standardization", *IEE Communications Magazine*, vol. 32, no. 1 (January), pp.
20–28.

Knight, Ivor, 1992, The Standards 'Summit' Meetings: Platform to the Global
Development of Users Telecommunications Requirements, paper presented at
the INTUG World Communications Seminar, Paris, February.

Lifchus, Ian M., 1985, "Standards Committee T1 - Telecommuni-cations",
IEEE Communications Magazine, vol. 23, no. 1 (January), pp. 34–37.

Onufrak, Joseph, 1992, Standards Strategies–Key to Profit or Loss? Paper
presented to the Pacific Telecommunications Conference, Honolulu, January.

Rose, Marshall T., 1990, *The Open Book: A Practical Perspective on* OSI, Prentice Hall, Englewood Cliffs, N.J.

Rutkowski, A.M., 1991, "Networking the Telecom Standards Bodies", *Connexions*, vol. 5, no. 9 (September), pp. 26–35.

———, 1994, Today's Cooperative Competitive Standards Environment for Pen Information and Telecommunications Networks and the Internet Standards-making Model, paper presented at the Standards Development and Information Infrastructure Workshop, Cambridge, Massachusetts, June.

Savage, James G., 1989, *The Politics of International Telecommuni-cations Regulation*, Westview Press, Boulder.

Schiller, Dan, and RosaLinda Fregoso, 1991, "A Private View of the Digital World", *Telecommunications Policy*, vol. 15, no. 3 (June), pp. 195–208.

Tarjanne, Pekka J., 1990, "Open Frameworks for Telecommunications in the 1990s: Access to Networks and Markets", *Telecommunications* (April), pp. 22–24, 48.

THE ENABLING POWER OF THE INTERNET: WINNERS AND LOSERS

MEHEROO JUSSAWALLA

1. INTRODUCTION

Those of us who live in the fascinating world of the Internet today never think of looking back to the time in 1844 when Samuel Morse exclaimed in wonderment "what hath God wrought" at the successful transmission of the world's first telegraph signals between Washington and Boston. Our frontiers are so expanded by the ease with which we access and transmit information that we rarely think of the price of the freedom to choose whatever information we want from the vast smorgasbord that has opened up before us through the technology of the Internet.

What is the price of freedom? When so many lives are lost and so much blood is being shed in many parts of the world? We might conceive this price to be very high. The Information Revolution will hopefully change all that since information is now both a commodity and a vital resource. The real price of acquiring it is steadily diminishing. What this means is that people in different parts of the world find themselves with many more options than their ancestors did and they are vested with the enabling power to make their own choices. The switches of information are passing out of the hands of the ruling elite and the large corporations as access to information is becoming democratized by the decline in its price. We are able to watch world events in real time everywhere and this explosion of information access via telephones, radio, television, and the print media have brought communities across continents in closer touch with each other. This transformation has led to the formation of rebellious majorities that have been demanding equity in access to information but have been denied this in the past. Their governments feel obliged to provide the facilities and the freedom of choice to obtain information in any form that the people want. In other words, the market for information has to be supplied with highways either by the public or the private sectors to enable two-way traffic to flow on them. The political network may remain local, but the information network is global and it transcends national frontiers. In the long run, it makes for peace and understanding among rival groups. In a society nurtured by competing multimedia sources, public choice expands in proportion to social needs for interaction. The regulators have been far outpaced by the dynamic technology which has brought households, corporations, banks, and even governments in closer collaboration than ever before.

The production, storage, transmission, and consumption of information is today mostly in digital form, rendering this activity a critical function in our civilization. The electronic ones and zeroes created by computers have reshaped our language in the post-industrial era. They crisscross the vast expanses of cyberspace and talk in a language that businesses and governments have all mastered in order to thrive in a competitive world. This mass digitization of information conduits is described as "virtualization" meaning that it recreates a virtual image of the real world. This has been the task of the Internet: to create virtual communities bringing people in an information exchange system without which they would never meet. It is unbound by barriers of space and time.

2.   THE EVOLUTION OF INTERNET

As far back as 1984 William Gibson created a futuristic world in his science fiction novel called *Neuromancer.* The world that he described was held together by a vast network of computers and telecommunications lines. He wrote that users of computers believed that there existed some kind of "actual space behind the screen" and he called such space "cyberspace." Today that cyberspace has become a reality and is termed as the Net, the Web, the Cloud, the Matrix, the Datasphere, the Electronic Frontier, and most recently, the Information Superhighway. Newt Gingrich describes cyberspace as the "land of knowledge" and "the exploration of that land as civilization's highest, truest calling."

Twenty years ago the Department of Defense created the Internet as a decentralized communications network that could withstand a nuclear attack. Then the National Science Foundation hooked into the network and made its supercomputer systems available to scientists and researchers. Universities were then connected to those centers. Today, the Internet is an intricate web of more than 39,000 networks linking an estimated 30 to 40 million users in 160 countries. It is estimated that every day about 1,000 new computers plug into the Internet which they consider to be the fast lane of the information superhighway. Anyone with a modem and a computer can access the Internet by signing on with either a public service provider or a paid service like America Online. No one owns the system and each network pays its own way. There is no single source to call if anything goes wrong because there is no set fee for Internet services. For some users, it is free, while others pay for the toll call to connect to a network, and still others pay service providers a flat monthly fee.

There are varied descriptions of the network. Anthony Rutkowski describes it as "a broad redefining paradigm," a transformation that builds the information infrastructure from the bottom up. The reason for its immense popularity is that it links innumerable computers on whatever telecommunications and computer platforms exist anywhere in the world. The problem it now faces is that its primary backbone, namely, the United States National Science Foundation, has decided to terminate it in 1995 and traffic will be distributed within the United States and around the world to other private backbone providers. Several new services have been developed because cyberspace has expanded to include millions of high-speed links to local area networks, E-mail systems fuelled by rapidly expanding wireless services. These are the means of transportation to cyberspace and not cyberspace itself. Cyberspace is about people who communicate with each other in electronic chat rooms and with bulletin boards and newsgroups. Negroponte (1995) defines it as an information world in which the fundamental particle is not the atom but the binary digit. The contents give the real value to the Internet. Buyers and sellers find each other without the expense of marketing campaigns. This is what makes the network a powerful engine of growth. In a world of haves and have-nots, the Internet allows everyone on the system to have equal rights and opportunities. Unlike traditional media, it is not a top-down model of communication. It does not use copyright software and is therefore open for anyone to use. No organization or government owns or controls it, and there are no laws to govern its usage. It covers various user groups like the Multi-User Dungeons (MUDs) and the Usenet with 10,000 discussion groups.

3. THE ADVANTAGES OF INTERNET

The greatest advantage of the Internet is that the content is created for consumers by consumers. This makes the medium more democratic and less hierarchical. The programming is not packaged in any American city and then broadcast to communities around the world, but rather is grassroots-oriented and as such is not considered a threat to the culture of any one nation or social order. Developing and develped countries permit its use by its citizens and business organizations even though a few, like Singapore, censure its users. Being unedited its content may be incorrect, boring, or a waste of time. Yet its users surf with software like Mosaic and Netscape and become addicted to its use.

The Internet is changing its content to make them more interesting and user-friendly by introducing "home pages" of the World Wide Web. The Web servers are scattered all over the world and contain every imaginable type of data. Software like Mosaic lets the user jump effortlessly from one Web computer to another creating the illusion of using one large computer. Hundreds of businesses, colleges, merchants, government agencies, and schools have been added to the list of Web servers. For example, the server from Paris provides information about the street maps, the metro routes, photos of city sights, and museum listings. The advantage to the user is that one can browse even on the Gopher. It is no longer a club for engineers and technophiles, but has become a medium for the masses. These changes in the use of the Internet are becoming quicker even as the technology of telecommunications boosts the speed with which data can move around the globe. As more fiber-optic cables replace copper, bandwidth expansion gives rise to the expectation of broadband fiber to the home. The National Information Infrastructure (NII) in America will be a broadband, switched network that will deliver information, distance learning programs, video phones, and television shows on demand. If these services are to become interactive and allow information from the home back into the network, even larger bandwidth will be required. This is so because users will become producers of information, the stage having been set by the Internet.

Technology as the enabler is rapidly changing the speed with which information on the Internet will soon be flowing. It is a supernet that does not offer all the services of Internet, but it does have the great advantage of speed. It is called the very high-speed backbone network service (vBNS) and is offered by MCI for the National Science Foundation. It is only for scientific use and not for browsing, and connects five research centers in the United States (San Diego, Pittsburg, Cornell, Boulder, and Urbana). Data travels at a speed of 155 million bits per second.

The objectives of Internet are its greatest advantages. They are to provide a seamless, friendly, affordable information network much like the telephone system developed in the 1930s. Internet has been greatly aided by long-distance telephone carriers such as Sprint, AT&T, and MCI that have laid fiber-based, high-speed telecommunication links that crisscross America, and their counterpart carriers do the same in other parts of the world. *The*

Economist (January 7, 1994) described the Internet as the "Grand Central Station of cyberspace" and the prototype of the information superhighway. It is estimated that within five years, there will be 100 million users logging on to the Internet daily. While it is paid for by American taxpayers and owned by no one in particular, countries around the globe have access to it. In the United States, commercial organizations pay monthly charges for their staff to send E-mail messages and browse through the net's three million files, downloading information and chatting with like-minded groups halfway around the globe at almost zero cost. Most universities and research organizations give their students and staff free Internet access.

Publishers are striking out on their own, using the Internet as their on-line kiosk. The success of such independent ventures offer advantages that the on-line services do not. The battle among on-line newsstands is becoming fiercer as Apple opened a service called eWorld and Microsoft will soon be opening the Marvel which will be bundled into every copy of Windows.

The advantage of Internet becoming user-friendly emerged from the invention of two systems: the World Wide Web and Mosaic. The latter allows the user to browse through the information in the Web in a point-and-click format. The advantage is that the page looks like it is nicely formatted as in a magazine and has a list of places and things. You can move from Geneva, Switzerland to the Yellow Pages of the Bay Area in California.

For large business corporations, the Internet has many advantages. Mary Cronin in her book *Doing Business on the Internet* states that there are short-term and long-term reasons for hooking up. But the most compelling reason is that the Internet is the biggest and earliest manifestation of the way business will be conducted from now on. Networked information is the standard for the future. Companies boast of increased productivity, better collaboration with strategic partners, and access to what is, in essence, the world's largest public library. Small companies are experimenting on the Internet with electronic advertising and large companies see the Internet as a potential marketplace.

James Gleick, author of *Chaos: Making a New Science*, which deals with the meaning and history of chaos theory, supplies businesses with a new service in New York city called the Pipeline. For them, the major benefits are E-mail. When an employee must be contacted from a hotel room in Pakistan, E-mail is safer and cheaper than the telephone. Businesses spend money to make their networks compatible with TCP/IP which is the Internet standard for addressing messages and sending data on the Internet. The result is that E-mail runs seamlesslessly inside and outside the company.

Collaboration between companies via the Internet is on the increase because of the major benefit of corporate R&D as well as exchange of graphics, video, and sharing of multimedia. Corporations find that the center

of the computing universe has shifted to a low-cost, powerful, open,
standards-based technology platform on which new products and services are
built. The future of computing is becoming defined by the Net. It has begun
to eclipse the PC and promises to wipe out the technical and geographic
hurdles that have been holding back the Information Revolution. Just as the
PC caused a revolution with its low-cost computing power, the Net brings
information and computing resources from all over the world to one's
fingertips. The boundary between one's own computer contents and the rest
of cyberspace is becoming imperceptible. From the user's point of view, there
is no difference whether a document is in a laboratory, the Library of
Congress, or the Web. It is a fresh field of dreams in which even small
companies can pioneer new technologies and set new standards (Verity and
Hof 1995). It makes no difference for the Net what software is used, what
chips are used by the user's personal computer, or in what language the
words are transmitted: the message gets through. It is the flexibility of the
Net that has made it appealing to users and has resulted in its rapid growth.
For example, an Israeli company called Vocaltec has delivered a $49 software
package that creates zero-cost, voice-messaging across the Internet. Thus
far, the voice links are not as good as the telephone but is likely to improve
with time.

4.   THE DISADVANTAGES ASSOCIATED WITH THE INTERNET

Some of the risks of shopping on E-mail on the Internet are those of revealing the user's credit card numbers or numbers of users' bank checking accounts. On February 15, 1995, Kevin Mitnick was arrested by the FBI for breaking into computers and pilfering thousands of credit card numbers from an Internet service provider. This is the kind of risk taken by consumers and corporations trying to do business on the Internet. Network-related crime is not new but with Internet being global, the opportunities have multiplied. A greater danger is one that involves the stealing of passwords and of employees determined to sabotage.

Corporations have learned to safeguard their information by building "firewalls" between their internal networks and the Internet. These are programs that screen incoming traffic so that only trusted computers can gain entry. Even so, spoofing is taking place in which hackers masquerade as trusted users. Software makers are now designing filters to prevent spoofing. Filters can block unauthorized outgoing messages. But all such protective devices are costly to design and install. To safeguard E-mail messages, credit card information, or sensitive data, encryption is superior to other devices. But so far most information sent over the Internet is unencrypted and thus vulnerable. Keeping unwanted intruders out of the Internet is not easy, but protection of information assets is gaining popularity.

Another problem emerging from the Internet is the legal aspects of libel which were ruled upon in the United States as recently as May 1995. In a New York state court, the judge ruled that Prodigy, a service supplier for computer communications through which the Internet can be accessed, was responsible for the messages on its network. An unknown Prodigy user had placed a series of messages on Internet alluding that a securities investment firm, Straton Oakmont of Long Island, New York, was fraudulent in its dealings and that its president was a criminal. Although Prodigy was only a conduit for the network, it was deemed responsible in the case. This has led to a bill being sponsored in the U.S. Congress to hold Internet's bulletin board and commercial information providers liable for the content of the messages sent through their systems. It is obvious that a conduit for communications cannot be held liable for the content of the information. The Internet is a worldwide system and laws differ in different countries. The troubling issue is that messages on the Internet can be anonymous and it becomes very difficult to pin down the conduit supplier for mischief done by any anonymous user. While it is true that Web surfers can view vivid pictures from the Louvre, the latest photographs from the Hubble telescope, or go shopping in virtual malls, society faces changes in its traditional notions of free speech and intellectual property rights that pose challenges.

With all the advantages gained from the Swiss invention of the Web, labor markets that are flexible, such as those in Silicon Valley, a science city, cannot be provided by the Internet. There are hundreds of jobs available to the science researcher in Silicon Valley but not to one working entirely on the Net in a virtual world. While the Internet provides decentralization, the larger companies concentrate in certain regions to take advantage of the labor market and this is something smaller and newer companies are still unable to do.

So long as the Internet was being used by a small club of computer virtuosos, the soul of the network was safe. But once the doors opened and a flood of users entered, the informal rules of self-policing became obsolete. Newcomers are generally shunned until they justify their existence on the net. There seems to be a clash between the mainstream culture and the hacker culture leading to an unpleasant war of words on the system. For purely technical reasons, it is impossible to censor the pornographic messages on the Internet. Likewise when the U.S. government put in the Clipper Chip, controversy raged about freedom of speech and communication among users. The government justified their action based on the need to intercept and decipher messages from terrorists, drug dealers, and the Mafia. Journalism has also suffered a blow from the publishers on the Internet inasmuch as the editor's task as a gatekeeper has almost disappeared. There is a marked shift in the paradigm of traditional journalism bringing in perhaps revolutionary changes.

Another significant disadvantage of the Internet is how does one overcome the problem of information overload or infobog as it is now known? The use of servers and filters may help to some extent. Today's technology has created the overload and there is no guarantee that future systems will help solve the problem. The knowledge economy is fundamentally different from the industrial economy and society must learn to come to terms with it. Users must learn to step back and gain some perspective on managing the deluge of information on one's screen.

One of the most confounding problems of the Internet is that while it has become global in its reach, it can only be accessed by the "haves" of most of the countries that use it. This is described as the most troubling aspect of the information age (Rattan 1995) because the new technology serves to widen the gap between the rich and poor people, not only in the industrially advanced countries but between the economically developed and developing ones as well. Access to information by a majority of persons in a given society will spell the success of democracy and in a computerized society, this will call for education and training of all its members and at all levels. What good is a National Information Infrastructure if it cannot be accessed by the entire nation? The same applies to a Global Infobahn if certain countries feel excluded and disenfranchised. It is true that market forces and competition are driving down the prices of computers and the paraphernalia required to access the Internet; even so these networks are available only in the metropolitan centers, and the rural areas in many countries are without a

dial-tone. Under such circumstances, it appears as if the new technologies

are breaking down society along traditional class divisions and this trend

must be stemmed by both the public and private sectors if democracy is to be

safeguarded.

5.   COST AND PRICE

So far the Internet backbone service has been provided at no cost to

users by the National Science Foundation in the United States. But in

December 1992, the NSF announced that this public good aspect of the net

will cease and that Internet will be commercialized. If free use of the ANS T3

backbone is stopped, then usage pricing will be required to keep the system

efficient. There are three competing backbones for the Internet. They are

first the ANSnet which is a nonprofit system that manages the publicly-

funded NSFNET for research and education, and also provides the backbone

for commercial users. The others are PSInet and Alternet which are

independent commercial providers of backbone services to both commercial

and noncommercial users. The technology for the Internet is quite different

from the one used for transmitting voice messages over telephone-switched

circuits. Unlike the Internet, in the packet-switching network, each circuit is

simultaneously shared by numerous users and the data can go by different

routes. As such, pricing models used for telephones cannot be used for data networks. The IP (Internet protocol) specifies how to break up data into packets, reassemble it, and send it to its final destination. The speed of this transmission has grown from 56 Kbps to 45 Mbps over the past five years.

Most of the cost of the Internet is a fixed cost, i.e., it is not dependent on usage. Thus, so long as the network is not saturated, the marginal or incremental cost of usage is zero. The NSF spends approximately $11.5 million per year to operate the net and provides a grant of $7 million to regional networks. The total cost is difficult to calculate because of the subsidies given by IBM and MCI, but it is estimated that the total cost has increased by a factor of 3.2 (Mackie-Mason and Varian 1993). Lease payments for communication lines and cost of routers form 80 percent of the total cost.

The costs of communication lines and routers have been declining rapidly at about 30 percent a year over the last three decades. In the past, communi-cation lines were cheaper than switches so that many lines fed into fewer switches. But the technology changed and now switches are cheaper than lines so that cost efficiency demanded that the data be broken into packets allowing many packets to flow over a single line.

Congestion on the Internet is another aspect of cost because it is not paid by the providers but by the users. When users must wait for a file transfer, it becomes a social cost. When there is congestion, the demand for bandwidth increases. Delay in delivery can be caused not only by congestion, but if a router fails then the packets are sent by different routes. This reflects the scarcity of the bandwidth in the network's delivery.

Faulhaber (1992) considered some of the economic issues related to pricing the Internet and suggested that transactions among institutions be based on capacity per unit of time. This implies that institutions would pay a monthly fee that varied with the size of the electronic pipe which carried the data. If the cost of providing the pipe was high, then the fee charged would also be high. This would apply to a dedicated line, which is a line connected to the Internet backbone. During times of congestion or shortage of bandwidth, the problem is one of figuring out transaction costs; but if the network is not congested, then the cost of sending additional packets is zero.

The economics of the system must recognize that the Internet uses scarce resources such as telecommunication lines, computer equipment, and labor. If these resources had not been used by Internet, they would have been used elsewhere. Since economics is concerned with allocating scarce resources among competing uses, Internet resources must also be so allocated. Even if the NSF were not to commercialize the Internet, it could still be priced. The goal is to find a pricing model that results in the most efficient allocation of the scarce resources being used. Pricing can be based on social value, which means that a packet that contains valuable information such as a video transmission of complicated surgery information to a remote region will be more useful than transmitting a video game. The problem is who will decide which packet is more valuable. Users will not want government to decide. In a pricing mechanism, the provider will announce the price to all users. Users then decide which packet is more useful to them and pay for them. When the system is congested and if prices reflect costs, then only packets with high value will be sent until the congestion is reduced.

The general criticism of the Internet is that low-income users are deprived of access. This is not a cost and price problem, but one of unequal distribution of wealth. This situation can be remedied by a system of subsidy if equity of access is to be assured. But on a global scale, each country will be responsible for providing access to its citizens at affordable costs.

6.   FUTURE PROSPECTS

In December 1993, Vice President Albert Gore in addressing the National Press Club in the United States predicted that "the changes coming in the related fields of telecommunications and computing are going together to make up one of the most powerful revolutions in the history of mankind" (*Wall Street Journal* 1993). Indeed the Internet with all its ramifications and new services has created the kind of two-way information superhighway that the Vice President had spoken about two years ago. A whole new information marketplace has been created in almost all countries of the world whether developed or developing. If the networks are not yet installed in the low-income countries, the declining prices and the determination of the people to join the global infobahn will force their governments and their private sectors to invest in building these bridges to the twenty-first century. Voice, video, and computers will be encircling the globe with enormous quantities of information travelling at faster speeds joined by the satellite skyways. With cellular mobile technology advancing rapidly and with the prospect of low earth orbiting satellites (LEOS) such as Iridium and Geodisic, it may become possible to access the Internet without the fiber optic networks that we use today. The prices of wireless communications are declining even as more and more users in developed and developing countries find uses for them in business and in their personal lives.

As the wave of deregulation and the spread of competition overtake Asia, Europe, and Latin America, the models initiated by the United States, the United Kingdom, and Japan are being overhauled. More regulators will find themselves taking a back seat as market forces work themselves out to the advantage of the users, giving the users greater choice in equipment and services at affordable costs. The marketplace of ideas will become more valuable giving opportunities for creative thinking. Simultaneously, there will be greater demands for protection of ideas and intellectual property which in the age of computer communications is most difficult to ensure. Already this has become a contentious issue between developed and developing countries who demand free flow of information and transfer of knowledge.

Just a short while ago, advertising on the Internet was unthinkable for Internet users based on the principle that on the Net one is expected to give as much as is received. A university researcher could offer a database program to colleagues in different countries for comments at no charge. Advertisers' messages are only one-way and are aimed at the mass market. It now appears as if advertising is taking over on the Internet without changing the culture of the network. Advertisers have been devising ways of placing their messages on the World Wide Web with attractive graphics and text, but not interfering with the bulletin boards that are an integral part of the network. But on the Internet there is no guarantee that the users will seek out the information provided by the advertisers. It is possible that in the future, advertisers will build virtual spaces to draw potential shoppers in large numbers. Already there are some companies in the United States that are creating electronic malls; examples include Esprit, Silicon Graphics, Open Market, Hello Direct, and MCI. Internauts browse through these at no charge. Those planning to travel are already using diverse vacation guides. The lure of advertising electronically go beyond the costs of printing and postage. It is the two-way communication between the supplier and the customer. Customers can send E-mail suggestions for product improvement or ask for technical support. Hello Direct has already placed its catalog on the Internet and updates it frequently. The prospect is that telephone sales will decline as the electronic mall grows and crosses national boundaries.

The value of the network increases exponentially with additional connections. Already Internauts are logging on from Bangkok to Brooklyn and *Business Week* in its November 14, 1994 issue predicted that there will be one hundred million machines in the next five years. Simple E-mail services on the Internet are already being used by seventy-five countries. It is expected that many will have full service in the near future.

From Arpanet to Internet, the most powerful innovation was the communications protocol IP which allows any number of networks to link up and act as one. Electronic publishers and private EDI suppliers are looking at Internet for low-cost alternatives. General Electric's EDI has been moving business forms between companies for over ten years at a high cost. This could be moved to the Internet at half the cost, but business corporations need assurance that their material will not be lost on the Internet or get delayed by congestion. Even broadcasters are experimenting with digitized audio and video snippets to PCs and this may be a trend of the future. Delivery of legal services to clients are also being tried out on the Net. As telecommunications circuits improve with digitization, it is possible that in the future, the Internet will blend in with the era of interactive television.

The Internet's unique technology and pricing scheme ensures that it will continue to grow in the future because its traffic can move over any channel be it a telephone line, cable TV, satellite links, fiber-optic trunks, or wireless phones. With the commercialization of the Internet, it is possible that telephone companies will take over its functions and will work towards providing security, privacy, and instant response. This provides a great opportunity for a growth industry specializing in innovative software and new services. There seem to be no bounds for the merger between human and artifical intelligence. The Internet will become the Global Information Superhighway which nations around the world are seeking to build.

REFERENCES

Business Week, *1995*, "Warding Off the Cyberspace Invaders", March 13, pp. 92–94.

Cronin, Mary J., 1994, *Doing Business On the Internet: How the Electronic Highway is Transforming American Companies*, Van Nostrand Reinhold, New York.

DeWitt, Philip Elmer, 1995, "Welcome to Cyberspace", *Time* (Spring special issue), pp. 4–8.

The Economist, 1995, "America's Information Highway", January 7, p. 35.

Faulhaber, G. R., 1992, "Pricing Internet: The Efficient Subsidy," in Kahin (ed.), *Building the Information Infrastructure*, McGraw Hill, New York.

Gibson, William, 1984, *Neuromancer*, Ace Science Fiction Books, New York.

Gleick, James, 1988, *Chaos: Making A New Science*, Viking Penguin, New York.

Mackie-Mason, Jeffrey, and Hal Varian, 1993, Pricing the Internet, paper presented to a conference on Public Access to the Internet, JFK School of Government, May.

Negroponte, Nicholas, 1995, *Being Digital*, A A. Knofp, New York.

Rattan, Suneel, 1995, "A New Divide Between Haves and Have-nots?" *Time* (Spring), p. 25.

Verity, John, and Robert Hof, 1995, "Planet Internet", *Business Week*, April 3, pp. 118–24.

Wall Street Journal, 1995, December 10.

A NEW ERA BEGINS:
TELEMEDIATION IN THE AGE OF PETABIT NETWORKS[†]

J. LOCKE

1.   INTRODUCTION

Internet, Information Super Highway, World Wide Web, HDTV, Direct Broadcast Satellite, Multimedia, Video On Demand...

The fusion of computers and telecommunications over the past twenty years has created a worldwide web of communication networks; networks that, while originally established for transmitting disparate content formats of either voice or data, have been gradually integrating their functionality and are rapidly being adopted by people who want to communicate with other people. Networks are not just technology or merely tools whereby we network; today's telecommunication networks have become the most widespread integrated communications system ever developed.

Over the past twenty years, technology has advanced to such an extent that access to telecommunications services has become virtually ubiquitous. Human communication has become the major use of telecomputing networks and is rapidly transforming these networks into a social space where people connect with one another. Technological changes, growing accessibility, and a dramatic increase in computational power continually leads to faster networks, new services, new uses, and, ultimately, new communities. This chapter discusses the history and near future of these newly emerging "telemediation networks" to illustrate differences and to emphasize the need for convergence, not only of technological innovations, but in the mental mind-sets of the leaders of disparate, yet converging industries.

The issues discussed in this chapter are meant to inform the converging worlds of telecommunications, broadcasting, and computing of the potential for opportunities and conflict by addressing such issues as:

- technological/functional innovations
- economic perspectives of the evolving telemediation industry
- divergent perspectives and models of information flow
- enabling technologies of the telemediation netscape
- future infrastructure needs and communication service designs

These issues are typical of those found across industry lines and are used to sketch the new "telemediation netscape" that will evolve from them; a netscape that crosses national, organizational, and time boundaries.

2.   WINDS OF CHANGE

Continuous improvements in both wired and wireless technologies promise a steep and steady decline in the cost of telecommunication services that is analogous to and parallels the way the reduction of the cost in computational power has continually fueled the computing revolution. More importantly, beneath the hype surrounding the recent convergence of mega-industries and their technologies lies a fundamental change in communication and the transfer of information. From initiation, acquisition, and transmission to delivery, the future road for communication and information processing appears to be following a common path; a path paved with the same materials, that is, sound, light, bits, and bytes.

Fiber optics technology dramatically improved the quality of telecommunications in the 1980s and is rapidly lighting the way to new dimensions. The entry of the computer into the world of telecommunications is facilitating the rapid convergence of new video, data, and image processing technologies; inextricably altering the industries from whence they came, while simultaneously reshaping and creating new ones. As narrowband telecommunication infrastructure capacities continue to be outstripped by emerging bandwidth-intensive applications and services, the confluence of broadband telecommunications and networking technologies has become a critical area for investigation. Advances in telecommunication technologies

such as fiber optics and asynchronous transfer mode switching (ATM)
continue to cross boundaries, bringing forth new ways to deliver voice
information and entertainment and uniting digital information flows with the
computer's evolving ability to simultaneously handle and display multiple,
simultaneous sessions of text, graphics, high-fidelity audio, and full-motion
imagery.

Claiming technological roots partially in each of the telecommunications, cable TV/broadcasting, and information processing industries,
"new media" and the digital technologies that make them possible are
revolutionizing the way in which many information-intensive communications
are accomplished. The implications these innovations hold independently for
the areas of information, media, and telecommunications are increased by an
order of magnitude when combined. This new synergy of digital acquisition,
processing, transmission, and delivery technologies opens the door to a whole
new world of applications and opportunities in business, communications,
education, and entertainment, one that I have come to regard as the "era of
telemediation."

2.1. The Global Marketplace

The proliferation of fast and cheap telecommunications has fostered the
expansion of what were once localized markets into truly global economies.
No longer are forces of change constrained to industrial or even national

boundaries. Dynamic innovations in information technologies, computing, telecommunications, and other support industries have accelerated the rate of change in these markets and are bringing about dramatic effects in the structure and organization of the information economy. The global efficiency of trade has increased as a result of optical fiber, undersea cables, satellite, and mobile communications technologies. The growing acceptance of innovations in signal switching that is capable of combining voice, video, and data capabilities over a single network are altering the traditional tele-communications landscape. The cost of existing telecommunications technologies continues to decline in the face of regulatory reform, while new technologies are destroying the natural monopolies of the carriers at an even more rapid pace.

2.1.1. Integrated Service and Global Access

The international standardization process allows for an open market approach to integrated network access. Standards for telecommunications support diverse services through integrated access arrangements while defining a limited set of standard, multipurpose interfaces for equipment vendors, network providers, and customers. Telephone networks around the world have been evolving toward the use of standardized digital transmission facilities and switches for many years. Only recently, however, has the vision of broadband telecommunication networks approached reality. In 1988, the

International Telegraph and Telephone Consultative Committee (CCITT) envisioned a universal network:

> a network, in general evolving from a telephony integrated digital network, that provides end-to-end digital connectivity to support a wide range of services, including voice and non-voice services, to which users have access by a limited set of multipurpose user-network interfaces. (from CCITT Recommendation I.110, 1988 in Kessler 1990)

The proliferation of broadband network technologies has fostered the international deployment of digital information technology while simultaneously creating a global environment ripe for expansion. This expansion is being motivated not only by technological advances, but by several market forces, including:

- a reduction in the cost of doing business due to telecomputing infrastructure investments and the advantages wrought by the support of universal services, portability of equipment, and reduced costs as standard chip sets become mass produced

- changes in management which reflects the demand for increased network simplicity, flexibility, and control over the evolution of infrastructure technologies as economical processes continue to push once widely divergent industries towards a common market

- improved capacity with the increasing availability of excess wired, wireless, and fiber capacities that can handle virtually unlimited amounts of traffic and new services with minimal or no capital investment

- the rise of global technology and information transfer that occurs long with the growing demand for domestic and international digital communications services for voice and data, particularly as businesses become more geographically dispersed and dependent upon transborder data flows, international coordination, and cooperation

  - growth of international and virtual markets as a result of the evolving demand for access to international markets for new services and products

  - technological leapfrog whereby for the first time in history, developing nations investing in telecommunications infrastructure development can "leapfrog" outdated information technologies and employ state-of-the-art facilities that compete directly with those of more developed nations.

Evolving broadband networking technologies and their associated standards are addressing many of these demands. However these advanced technologies also promise a major upheaval of the traditional "telecommunications landscape."

2.2. Evolution of the Telemediation Netscape:

Open Systems Integration

In the current state of data and telecommunications, users must rely upon different physical and logical interfaces to interconnect disparate, task-specific networks. In regards to telephony, for example, a user accessing a public telephone network requires a connection to a central office and telephone equipment that follows the correct protocols for accessing the network. In terms of data communications, the user who wants to access a high-speed broadband (^{3}56k bit/s) public network will require an additional set of wires from the same central office or a packet network provider and will employ an additional set of hardware following the correct protocol for packetized data services (X.25 etc.).

As new media services proliferate, the same issues of dedicated access and hardware-based connectivity protocols will continue to be problematic. Cable television, having evolved beyond its original goal of community antennae television, is technologically capable of offering limited interactive services; but to do so requires yet another physical interface (a coaxial cable) and a hardware device that can convert the cable signal to one that a television can understand. Utilizing its own coaxial network and/or a combination of both the cable and telephony networks, interactivity is provided by multiple disparate devices and communication networks. As new media applications and services rise in popularity, each new service and provider under the current infrastructure will require an additional communications path, different protocols, and different network facilities.

Internal communication developments coupled with advancing device development are altering these technological roadblocks. A closer examination of the demands new media services will place upon the existing infrastructure can offer clues as to what services and technologies will be made available to businesses and households of the future. Above all else, one point is glaringly obvious. Today's POTS connections will be wholly insufficient for the next generation of communications services. While ISDN appears to be the best suited of the potential telephony-based telecommunications architectures to address the growing problems of multiprotocol network and hardware dependencies in the local loop, years of delay in the implementation of ISDN has brought upon the telecommuni-cations industry strong competition. The growth in penetration rates of cable television in the homes and data network providers in the corporate world, and the proliferation of independent networks have given strong footholds to a new breed of competitors in the telecommunications service industry.

2.3. The Telemediator Revolution: Fall of the Dedicated Box

The consumer electronics, communications, and computer industries are on the verge of delivering a host of new products and services to the consumer marketplace over the next decade. Few industries have

experienced the growth and confluence of capabilities that these once wholly
divergent industries have. The market shares of high-cost computational
hardware such as mainframes, digital signaling units (DSUs) terminals, and
dedicated transaction processors have declined steadily in the face of stiff
competition from smaller yet more capable mini- and microcomputers. The
tendency to downsize and distribute central processing capacities has hit the
telecommunications industry as well. Dedicated hardware devices such as
digital signal processors that were expensive have been reduced to a
microchip. Telecommunications capabilities are encoded in software. Laptop
computers with built-in modems and data networking capabilities have
become commodities.

The 1990s are witness to the greatest amalgamation ever. For some
time, there has been a gradual shift from format, protocol, and task-specific
devices to microprocessor-based, digital systems that allow a much greater
degree of flexibility (and cost savings) over dedicated analog devices. The
model of a format independent network of services and interface devices, such
as the CCITT's Integrated Services Digital Network recommendation, is very
attractive to industries that have been previously hampered by format wars,
declining revenues, and increased uncertainty as to the correct path towards
an era of broadband telemediation networks.

The whole concept of an open digital information processing systems replacing task-specific and proprietary systems has been hotly debated for the past several years in many industries. Even in the telecommunications industry, manufacturers of "traditional" telecommunications switching and computational hardware are split on the issue. No industry, however, has attacked these issues with more resources or vigor than the computer data communications industry. While some manufacturers choose only to market solution-specific hardware devices, others are creating partnerships with hardware and software manufacturers to blend expertise in application-specific LAN to WAN network interfaces, user interface designs, networking protocols, and a knowledge of end-user market requirements with the computer manufacturer's computational engines, open operating systems, networking capabilities, and access to a world of software developers who make it their business to produce marketable and successful hardware and software solutions to the home and corporate markets. The result of this metamorphosis is a new class of information processing and telecommuni-cation devices which I refer to as "telemediators" (telecommunications capable multimedia information processors).

Telemediation is increasingly used by a wide variety of different people for a variety of tasks around the world. Over the past twenty years, technology has advanced to such an extent that almost everyone comes in contact with the traditional "computer" in one way or another. Unlike in the early days of mainframe computing, when only highly skilled technical people could use computers, the range and experience of users today continues to increase as personal computing systems become ubiquitous, intuitive, affordable, accessible, and easy to use. Even today's answering machines are minicomputers, utilizing application-specific integrated circuits (ASICS) to handle all message functions and store incoming messages.

The next generation of devices to support telemediation, however, is extremely powerful and is rapidly pushing the boundaries of previously stable industries. The question that leaders of the converging cable, telecommunications, and computer industries should be asking themselves is what functions can and will the telemediator of the future provide? It is well known that telemediators are no longer single-function machines, but provide superior quality, digital voice and video capabilities in addition to increasing computational capacities. Cutting-edge visual capabilities include digital optics and compositing for film, 3D animation (Jurassic Park), on-air interactive graphics (the BBC's coverage of Clinton's election), and film restoration (Snow White and the Seven Dwarfs). These devices are capable of far more than low-quality video digitization technologies that are offered by MPEG, Quicktime Video, and Intel DVI today.

Interactive capabilities combined with data communications are giving consumers the ability to conduct from a desktop remote monitoring, gaming, virtual reality simulations, electronic document interchange, file transfers, and application sharing. Many of the tasks associated with telecommunications can easily be handled by telemediation devices and software including digital audio sampling and transmission at 16bit /44.1 kHz, packet-based video conferencing and routing (CUSeeMe from Cornell University), PC-based Video Phone, and, most importantly, global network connectivity.

Herein lies the heart of the debate. As solution-specific devices and networks continue to be replaced by general purpose telemediation systems, how will the newly emerging new media markets be affected?

2.3.1. New Concepts in Telemediation

For multimedia information processing, the last year has seen an explosion of alternative applications and devices. The stereotypical audiovisual capable PC is one that uses asymmetrical compression codecs to facilitate the handling of audio and video streams to and from local disk and memory. Such systems are often capable of a few proprietary file formats and are generally designed for use as stand-alone systems. This class of product has been very quick to catch on and is one of the earliest and most visible examples of the trend towards open architecture, multiple purpose, digital media systems. The telemediator takes this class of service and product to a higher level.

For global telemediation, the ability to handle, from a personal desktop system, long-distance voice and video information in real time along with traditional computer applications provides a distinct and separate choice from traditional telecommunications service alternatives. The current provider of these services, however, is not your local telecommunications provider, but rather a global data network of more than twenty million users including commercial, university, research laboratory, and nonprofit agencies, i.e., the Internet.

Those unaware of the information processing revolution being spearheaded by global access to the Internet should take heed. Corporations who ordinarily rely on their leased-line, private data networks to cut telecommunications charges rather than over the public telephone network now have another option. For others with low calling volumes, where leased lines are inefficient and not cost effective, the global information access the Internet provides is also attractive.

2.4. Telecommunication Policy and Deregulation:

The Economics Of Resource Allocation

In a perfect world, access to telecommunications services would be ubiquitous. Every telecommunications network would be run at maximum capacity for 24 hours a day, yielding the greatest return on investment in the infrastructure. Services would be global in scope allowing for shifting traffic patterns along time zones and across international borders. However, with the current global telecommunications landscape, this could never happen. The reason for this, of course, lies in the regulatory bodies of both developed and emerging nations and the continued proliferation of voice-based narrowband standards.

U.S. regulatory policy is based on the notion of telecommunications services as regulated monopolies. This assumes that due to the vast nature and cost of infrastructure construction, only one system could be built. Historically, such a policy set a precedent of granting monopoly rights to service providers in exchange for regulated rates and universal access rights for subscribers. For their time (about 1934), these concessions seemed appropriate considering the cost and difficulty of such an undertaking.

Since the breakup of AT&T in 1984, however, such political philosophies have been criticized as being outdated and protectionist. The International Telecommunications Union (ITU) notes that in every market where competition has been introduced, rates have fallen and calling volumes have risen dramatically. While it has been reported that between 1984 and 1994 AT&T lost as much as 40 percent of its market share due to competition, in the same period, it has dropped its rates by 60 percent maintaining operating profits of $4.71 billion on revenues of $75.09 billion.

This represents a 26 percent increase over its pre-breakup performance on returns in 1984 (Arnst 1995).

As policymakers attempt to navigate through the ever-changing seas of innovation and economic change, a deregulatory movement has begun to circle the globe. A new flood of players and services is pouring into the rapidly deregulating U.S. telecommunications market. Existing and potential service providers are spending billions of dollars to create stronger brand names and build broadband networks. Running to meet the call of cable providers and other competitors, Pacific Telesis among others has recently announced plans to spend up to $16 billion by the year 2002 to upgrade its telecommunications infrastructure. Other providers are investing heavily as well: for example, Bell Atlantic is investing $11 billion; U.S. West, $10 billion; and Southern New England Telecommunications, $4.4 billion.

Market analysts, KPMG Peat Marwick, estimate that in 1994 alone worldwide mergers and deals among communication, information, and entertainment companies reached $27.8 billion. Some of the heavy speculators in this rapidly deregulating field include:

- AT&T which invested $12.7 billion for ownership or controlling stakes in McCaw Cellular; 3DO, Spectrum Holobyte (Games); Knowledge Adventure (Software Publishing); and General Magic (Software For Personal Link Massaging System)

- TeleCommunications Inc., Cox Enterprise, and Comcast which formed an alliance with Sprint USA. Regulation reform permitting, the group intends to offer one-stop shopping for local calling, long-distance calling, cellular, PCs, and cable TV

- Pacific Telesis, Bell Atlantic, and NYNEX formed a joint venture with Hollywood's Creative Artists Agency to develop new media content

- Ameritech, Bell South, and SBC formed a joint venture with Disney to develop new media content

- West formed a joint venture with Time/Warner to develop new media content

- NYNEX spent $1.2 billion in a joint venture with VIACOM to develop new media content.

Why such haste? The economics which made the natural monopoly a necessity no longer exists and is rapidly giving way to free market forces. Even the once "sacred" Bell Communications Research Laboratories has recently been divested and put up for sale by its owners, the regional Bell operating companies (RBOCs). The $1 billion company has been known for developing technical standards and was originally established as a research facility for the RBOCs, all of whom share a common lineage to AT&T. But as the RBOCs diversify their services, they face increasing competition from one another as well as from other converging industries and they no longer wish to cooperate in funding Bellcore. Further precedents of deregulation such as the granting of permission for Bell Atlantic to offer video-dial tone services in Arlington, Virginia and pending decisions regarding cable TV based on local loop competition has set the tone for a mega-convergence of multitrillion dollar industries.

Just how each industry giant plans to capture its share of the emerging telemediation market is reflected by the organizational cultures they bring to the table and is discussed further in this chapter. At stake is competitive access to several lucrative global markets and industries including telecommunications, publishing, entertainment, education, broadcasting, personal computing, and consumer electronics.

Varying perspectives call for different service and provider plans; however, central to these new services is a broadband-capable networking infrastructure. Global broadband network resources such as optical media and the emerging ATM switching standard can be reallocated to new services simply by launching a different software configuration. These same networks will be capable of handling multiple sessions of synchronous, isochronous, and asynchronous data, voice, and video communications over the same infrastructure of fiber optics, permanent, and switched virtual circuits. It is with these tools that the future of telemediation is being forged.

3.   DIVERGENT PERSPECTIVES OF INFORMATION FLOW: CLOSING THE TELEMEDIATION GAP

In many ways, watching the New Media Revolution is like watching the rebirth of the software industry in the early 1980s. With the introduction and rapid diffusion of the personal computer, software developers had to divert resources from their traditional large-scale mainframe paradigm and learn new ways to manipulate digital data. They had to invent new roles, new paradigms, new interfaces for both man and machine, new forms of interaction, new processes, new products, and new techniques for bringing them to market.

Today, telecommunications providers are facing the same challenge. In the face of competitive-access regulation, the task of clarifying the roles and players of the emerging telemediation industry is falling upon a collective of both service and consumer product industries. Vendors, distributors, and information providers in publishing, music, art, and education are allying themselves with consumer electronics, computer, office equipment, and publishing houses to stake a claim in this volatile and rapidly evolving frontier. The industries to be most affected by this digital boom, however, are the telecommunications, data networking, and broadcasting industries which share little common history. Suddenly, these industries are being faced with direct competition from a variety of widely divergent yet equally powerful players. How they approach these developing markets is inextricably bound to their organizational culture and the industrial paradigms from which they emerged.

3.1. The Broadcast Paradigm

One of the oldest and most powerful of all industries to participate in the new media revolution is that of broadcasting. Since the advent of Marconi's "wireless" in the early 1900s, broadcasting has enjoyed a highly visible and influential position among consumers. Driven by advertising sales, the broadcasting industry for the last half century has derived its revenues from essentially providing an information service to the public.

In its simplest manifestation, the paradigm best suited to illuminate the underlying industrial culture of broadcasting is reflected in the Information Transmission Model (Shannon and Weaver 1949). Viewing the consumer as a passive receiver, information (i.e., entertainment, news, advertising) is transmitted through a specific channel to a receiver, where it is passively absorbed at the final destination by the consumer. Emphasis has traditionally been placed on the one-way transmission of a message or programming, upon which the consumer was expected to respond by purchasing the advertised products carried over the channel.

The evolution of the broadcasting network infrastructure is a reflection of the information transmission model. Adapted later by television and eventually community antennae television (cable TV), little has changed. Historically speaking, informational content has been transmitted from a central source over either wired or wireless infrastructures, where it is later repeated, distributed, and ultimately delivered to the receiver. Under such conditions, information provided by the broadcasting industry has always been unidirectional, asynchronous, content-specific, static, and pre-defined. Yet this model and mechanism for sharing distributed information has been extremely effective and influential overall.

For those participants of the new media revolution evolving from the broadcast paradigm, the hype over new media appears more of an evolution than a revolution of their traditional industry. Continually based upon a product-driven market of distributed information or programming, the new media age is perceived as merely an expansion of existing patterns of information delivery to include text, images, audio, and full-motion imagery over additional physical mediums. The future market remains product-driven, where deliverable goods are passive and pre-defined data are supplied by remote servers acting as proxies on behalf of the content creators.

3.2. The Telecommunications Paradigm

The telecommunications industry can be said to have developed from a "limited conversational model." Driven by providing interactive communication services over a peer-to-peer infrastructure, the limited conversational approach allows for the natural flow of dynamically determined content, information, or data such as voice. Essentially, the facilitation of real-time interactive communication has been the backbone of the telecommunications industry.

As technologies advance, the telecommunications industry has been reluctant to incorporate the latest innovations into their services. As a natural monopoly, there were no threats of serious competition from other industries and, thus, little motivation for technological innovation in services provided to the consumer. The telecommunications industry has historically catered to a service-driven market, whereby revenues were derived by providing access to the network. Since their inception, there have been few alternatives for point-to-point telecommunications that were not soon dominated and controlled by the major forces in the telecommunications industry.

For those participants of the new media revolution evolving from the telecommunications paradigm, the evolution of new media services and competitive providers will have a more significant impact. Telecommunications providers today are spending billions of dollars to upgrade their aged and insufficient analog networks with digital and optical technologies. Since the introduction of competition following the breakup of AT&T in 1984, long-distance communications have improved and have been facilitated by the conversion to digital switching and fiber optics. Additional interactive service plans are evolving to augment traditional point-to-point communication services and local and long-distance calling to include mobile, cellular, paging, and PCs. Ultimately, many of the telecommunications providers aspire to expand their role in the era of telemediation by becoming content providers as well as competitive access providers to the underlying network infrastructure.

The technological infrastructure of the telecommunications industry is better suited to handle the pending transition to telemediated services than those of competing industries. Already in place are the physical connections, switching, control, and tracking mechanisms needed for point-to-point transactions. These advantages carry with them critical inadequacies, however, which could have a profound effect upon the future of telecommunications-based services. No longer can telephony alone sustain the industry in a competitive market.

The most significant challenge facing the industry today is bandwidth. Whereas broadcast and cable infrastructures currently lack the point-to-point and switching capabilities of today's telecommunications networks, they have

a major advantage in terms of the potential bandwidth capacity available to

consumers. By nature of being a competitive industry, heavy investments

have been made over the last decade in the modification of existing network

infrastructures to include fiber optic transmission plants from the facility to

the neighborhood distribution points. The "last mile," connecting fiber

backbone transmission plants to the home, continues to be copper-based.

Unlike the Plain Old Telephone System (POTS), however, this connection is

provided by coaxial cable, a medium that is sufficiently capable of carrying

the high bandwidth services that the new media will demand. The resulting

competitive danger to the telecommunications industry is that with only

minor modifications to the existing infrastructure of cable and broadcast

networks, broadband telecommunications services will become a reality.

The network infrastructure of the telecommunications providers

remains overladen with antiquated analog and copper technologies and could

ultimately overshadow aspirations by the telco's to thwart competitive access

providers with more advanced network capabilities.

3.3. Evolution of the Telecast Paradigm: A Limited Experience Port

For telecommunications and competing industries, new media markets

will become increasingly product- and service-driven. Revenues will be

generated by providing competitive access to a network infrastructure

233

sufficiently capable of carrying high-bandwidth new media services, as well as providing the content for such services. Most of the investment by new media companies today appears to be placed in the construction of fiber/coax systems which closely parallels the client/server networking paradigm of the computer software industry.

Interactive Television trials such as Time Warner's Full Service Network in Orlando, Florida exemplify this perception. After spending hundreds of millions of dollars to implement a high-bandwidth, fiber/coax infrastructure and to integrate numerous untested technologies, Time Warner intends to deliver new interactive services to the home. Yet return bandwidth is limited to under 1 Mbit/s, a data rate that alternative technologies over copper cannot even approach at this time. Furthermore, content remains deliverable with pre-defined data supplied by remote servers acting as proxies on behalf of the content creators. Such a vision does not do justice to the potential of telemediated services.

Through hybrid fiber/coax frequency division multiplexed networks envisioned by telecommunications and cable/broadcast providers, service providers of the new media age initially intend to offer subscribers a "limited port" experience. Paralleling the client/server computational architecture, clients (set-top interfaces) will provide users with a high-resolution, 3D user interface which performs all necessary digital media processing and request brokering through a low bandwidth return channel. At the head end or central switch, a server maintains several databases and processes requests from the client to extract data (video, home shopping services, etc.) from the database. It is through such systems that new media providers envision delivering innovative services such as media on demand, home shopping, banking, interactive programming, remote application control services, informational services, telecommunication services, answering service, call forwarding, and paging.

Electronic commerce, entertainment, and information are perceived to be the wave of the future and the driving force behind capital expansion and investment in a market that is increasingly being viewed as product-driven. The industrial strategies evolving to address this vision has led to the emergence of what I refer to as a "telecast paradigm," an approach which differs little from the long-standing perspectives of the broadcast industry.

If the ultimate goal of telecommunications and cable operators is to support the vision of the telecast paradigm, CATV networks have an apparent advantage over telephony networks in that existing coaxial and emerging fiber/coax infrastructures have been optimized for one-way distribution of TV signals. The most efficient frequency plan allocated only a small amount of bandwidth, in the 5–30 MHz range, for return transmissions to the head end, while the bulk of the spectrum (54 MHz to 750 MHz) has been allocated for downstream communications. New data compression technologies will allow from four to ten channels of digital video to be sent over the same bandwidth (6MHz) that is required for a single analog channel. Emerging fiber/coax systems typically offer seventy analog channels and several hundred digital channels (add a switch and some number of switched digital channels can also be delivered). Potentially, allocations of unused discreet frequencies within the network infrastructure could be made available for multiple services.

The relative value of upstream communications, as perceived by the telecommunications and cable operators promoting the telecast paradigm, is reflected in the proposed interactive services of either industry: broadband, one-way scheduled digital video data broadcasts, point-to-multipoint "on demand" services with narrowband return, or narrowband two-way audio/data communications. Virtually all major players are envisioning services of the limited experience port model, through which the subscriber has little more than a limited port (low bandwidth) of interactivity. Some of the more innovative service plans include higher bandwidth return allocations through the use of two disparate networks including LEC and cable infrastructures; however even under these plans, users may only retrieve pre-determined, static information with limited control over delivery time and accompanying start/stop capacities.

Is the future of interactive broadband telecommunications to be dictated by such a narrow view of interactivity? Is the investment of billions of dollars in public infrastructure to support this vision warranted? The new full service networks appear to be built upon a "field of dreams," that is, with the assumption that "if you build it they will come." Providers of these new on-demand services foresee becoming an integrated communications company, ultimately offering local calling, long distance, cellular, PCS, cable, and interactive audio/video applications, or becoming content creators and providers. Unfortunately for the consumer, by limiting their perspectives to an ideal which can be depicted by the statement "father knows best" (large pipeline for downstream data alongside a limited control only "interactive" return signal), this vision hardly seems credible. For services to support truly interactive applications, they must support the traditional peer-to-peer architecture that telecommunications companies were founded upon; that is, symmetrical broadband services must be provided equally on behalf of providers and end users.

3.4. Data Communications/Networking Paradigm

While technical innovations such as Time Warner's "Fiber Deep" architecture could support symmetrical bandwidth demands sufficient to compete in the LEC market, there are numerous other conceivable demands for bandwidth spectrum. If the broadcast/cable and telecommunications companies hedge on the infrastructure costs of a systemwide upgrade allowing real-time, symmetric, dynamically switched, and controlled peer-to-peer broadband telemediation, there is an industry ready and waiting to step in: data communication networking providers. Their goal is to create an environment in which users can consult and collaborate with one another with the ease of a telephone call and include in the conversation the active use of their own applications, information, and data including full bandwidth voice and bidirectional video.

The market data networking access that providers currently cater to is primarily reserved for the corporate world; that is, a service-driven, completely digital marketplace where producers and users of information are brought together in real time, with respective sharing of tools and data to support such operations as the "virtual office" and ad hoc telecollaborative meetings. Many of these interactive applications have been pioneered in the local area networking environment and are slow to be recognized by LEC and cable operators.

For the data networking industry, service is based upon a fully functional collaboration/conversational model rather than the limited experience port model envisioned by many cable and telecommunications providers. Such services are the most demanding in terms of bandwidth and computational power, where facilitation of highly interactive telemediation is critical. Service requirements demand a highly interactive peer-to-peer networking environment with real-time bidirectional synchronous interaction among users who work together to review and produce information, while simultaneously and transparently providing for natural multidimensional and multimodal data flows including speaking, visual collaboration, gesturing, and data sharing.

The data communications industry for years has been dwarfed in terms of revenues by the telecommunications and broadcasting industries. Yet as the digital boom continues, it is the data communications industry that appears the best prepared to deal with the volatile market demands and dramatic economic implications of the evolving telemediation industry. A competitive industry recovering from major cutbacks and recession during the late 1980s and early 1990s, data communications providers have already suffered the pains of market competition; something both the cable and telecommunications industries will have to face shortly.

Historically, limited penetration of costly computational hardware and incomplete network infrastructures hampered the data communication industry with limited functionality and a much smaller market share than the large telecommunications providers. This is no longer the case. Already, IBM operates the world's largest data network and resells voice services over excess capacity in its state-of-the art optical fiber infrastructure. Continuous improvements in synchronous optical transmission technology have increased the capacity of a single fiber from 32,000 to 320,000 (64 kb/s) digital transmission channels. And while IBM has publicly stated that it does not intend to directly compete with voice carriers, it will sell customer services so as to conduct electronic commerce.

Moreover, the decreasing cost of computational hardware is putting devices capable of supporting multichannel broadband digital telecommunications access in the home. One of these devices is the "smart" set-top boxes currently being promoted by computer and consumer electronics manufacturers such as Pioneer Electronics, General Instruments, Scientific Atlanta, and Hewlett Packard. Recent interactive trials of the Full Service Network concept in Florida by Time Warner have utilized a set-top box created by Silicon Graphics Incorporated as the interface device for their media on demand architecture. In its current configuration, this device is limited to receiving and decoding digital information (MPEG I; II movies) in real time for subscribers of Time Warner's media on demand services.

At first, the use of another set-top box may appear to be a simple, technology-specific modification over existing cable tuners. However, in reality, they represent a paradigm shift in the delivery of information and digital media processing. The potential of these new interface technologies is hidden by their apparent limited functionality.

The "set-top box" of the future is equivalent to a $20,000 workstation computer today. Computational performance over the past twenty years has enjoyed a factorial price/improvement factor of ten every three years (as measured by throughput, a combination of internal processing speed, I/O speeds, and the efficiency of the operating system and other system software working together (see Birkmaier 1993). Where initially large rooms were filled with cases of transistors and tubes were chained together to run data processing programs entered on "punch cards," today's computers have eclipsed such primitive technologies and continue to revolutionize the information processing industry.

The once venerable Cray Super Computer—a 75 MHz, 64-bit machine with a peak processing power of 160 Megaflops—was thought to be the ultimate computing machine when it was developed. In 1976, the Cray 1 had a price of several million dollars. Today, you can purchase a microchip-based, parallel processing (2 CPU, 150 MHz/64 bit) machine (Silicon Graphics Challenge L) with a peak processing power of 600 MFlops for about $250,000; add an additional 16 CPUs (and a lot of money) and you can increase that throughput to 5.4 Gflops. That is 5.4 billion floating point (1,0n) calculations per second!

3.5. The Telemediation Netscape

What, you may ask, does all this about computational power have to do with telecommunications, broadcasting, and the new media markets? New media and the accompanying interactive service capabilities are essentially digital, and computational power that is provided by supercomputers is the cornerstone upon which the industry is being built. Only supercomputers have the computing power necessary to handle the vast amounts of data required for digital video and other media on demand services. Interactive networks must deliver information at the same time it looks for and interprets input from the user. In the case of today's proposed interactive networks, where the head end will be expected to provide the consumer with interactive capabilities and services, networking control systems must be able to simultaneously respond to messages sent through the network and handle thousands of users.

Consider the Time Warner trials mentioned earlier. Each of the technologies responsible for accepting, transmitting, compressing, decompressing, rendering, and managing the delivery of voice, text, video, and real-time graphics is digital. At the receiving end is a set-top box with the power to decode digital video "on the fly" while simultaneously providing access to other user services. Specifically, each set-top box provided by Silicon Graphics Inc. for the FSN trials contains a system capable of digital video and audio processing, multiple session management (multitasking), and high-speed data communications (100 MHz R4600 RISC processor, 8MB RAM, 24-bit color graphics, and multiple serial, parallel, ISDN, and Ethernet capable networking interfaces). While not as powerful as the supercomputer serving the information from the head end facility, this system still performs at a rating of 25 million floating point optical computations per second.

Of greater significance to alternative access providers is the fact that many of these boxes have multiple I/O ports accessible to the user, including 8-way RJ45 for T-1, E-1 and RS422 connections through ordinary phone lines, threaded F-receptacles for coaxial cable connections, sub-DIN RS-232 connections for PC connectivity and remote device control, 21-pin EURO-AV for SCART interface connection to European PAL sets, mini-DIN for S-VHS, and RCA connectors for composite video and audio connections. The keyboard and high-resolution monitors that accompany workstations have been replaced by a limited function remote control, an interface, and your television set.

Initially, consumers will lease the first wave of set-top boxes, but models and advanced features will eventually be sold through retail outlets. After-market add ons could include built-in, high-density digital video disk drives that share the MPEG II circuitry with the set-top box. At a final projected cost of $200–400 per interface, such computational power ten years ago would have been a fiscal impossibility. However, the labor cost of developing a computational set-top system is a constant just like the workstation counterpart. When combined with the increase in capacities and the potential market of over sixty million units, the projected costs plummet.

What effect will the digital boom have upon the industries competing for access to the home and corporate telecommunications markets? When the set-top boxes can be placed in a variety of proposed interactive networks and perform a wide variety of functions, the fundamental question becomes how will interactive telemediation services enter homes and businesses? Will it be done through cable, copper wire, or by radio waves broadcast to a satellite receiver? Telemediation services are much more demanding and bandwidth-intensive than voice alone. Emerging full bandwidth-interactive applications such as desktop video conferencing and data sharing, while not in direct competition with voice services at this time, provide a much richer form of point-to-point interaction than telephony. Given the choice, consumers may prefer to change information carriers to those who are capable of offering the services they desire. Existing data communications architectures hold a competitive advantage to support these services by enabling symmetrical, end-to-end digital transmission and connections. Through integrated access arrangements with competitive access providers, these network operators can offer diverse services, internetworking individuals across publicly switched telephone networks and cable network infrastructures. Broadband digital networks such as those operated by IBM can handle virtually unlimited amounts of traffic regardless of media type with minimal or no new capital investment. In an open market, these capabilities pose a very real threat to telecommunications and broadcasting giants...has anyone heard the story of David and Goliath?

4. THE TELEMEDIATION INFRASTRUCTURE: ENABLING TECHNOLOGIES FOR HIGH BANDWIDTH APPLICATIONS

Increasingly, regulatory reform, internal communication, and device developments have accompanied the integration of physical devices laying the foundation for what I refer to as the telemediation netscape. Competition between telecommunications and cable/broadcasting has become a very real concern to both industries. Regulatory barriers are falling while new architectures and technologies are making the goal of high-bandwidth, interactive telecommunications a reality.

Transmission capacities have mushroomed with the proliferation of multiple transport options such as fiber to the curb, hybrid fiber/coax, and even copper-based asymmetrical digital subscriber line technologies. Fiber/coax systems capable of carrying voice and data on the same plant have evolved along with switching technologies (ATM) that will packetize, address, and deliver multiprotocol data streams including TCP/IP, MPEG video, and PCM audio to individual set-top boxes. The developments in broadband transmission, switching, and processing capacities; fiber optics; micro-processing power; media compression; and network intelligence have enabled us to build a variety of networks for new media applications and services using the different strategies, perspectives, and models addressed in this chapter. Each of these technological innovations is contributing to a paradigm shift and deserves closer examination.

4.1. Broadband Transport and Switching

The noteworthy movement of the 1990s is the penetration of optical media communications from backbones to businesses and eventually to the home. Fiber optics technology dramatically improved the quality of long-distance voice communication in the 1980s and is rapidly leading the way to new dimensions. As the 300–3000Hz signal was digitized, multiplexed, regenerated, and error-corrected, artifacts from the analog era including hums, hisses, crackles, pops, distortion, and crosstalk were eliminated. Combined with the ability to transport any type of digital media information and to switch that information over optical circuits, whether virtual or permanent, optical transmission and switching technologies have become the de facto standards for the new media age.

4.1.1. Optical Media Transport

Fiber's signal transmission characteristics are dramatically better than those of copper. In a fiber-based telecommunications infrastructure, a single voice signal is transmitted at 56Kb/s, a rate designated at DS-0 (Digital Signaling-0). But no signal travels alone once in the digital domain. Twenty-eight of them, Time Division Multiplexed (TDM'd) together after digitizing, are designated as a DS-1 group (1.544 Mb/s). When, for efficiency sake, twenty-four DS-1 Groups are TDM'd together, the resulting DS-3 (44.7Mb/s) signal brings together 672 individual voice signal transmissions into a chorus of light. These two groupings form the foundation of many telecommunications networks today and are widely tariffed for use by both telecommunications and data communications industries.

The telemediation netscape that is required for tomorrow's broadband services must be able to handle rich, full-motion media that have the distinct ability to capture the imagination of the subscriber, dynamically and in real time. Broadband networking infrastructures must be able to handle a variety of switched and nonswitched circuits allowing flexible throughput of multiple communications media, including synchronous, asynchronous, and isochronous data and information types. Compared to the multiplicity of copper-twisted, pair, or coaxial cables that are required to transmit today's television video, audio, and data signal groups, a single fiber can carry dozens of multiplexed signals, translating into enormous hardware and installation labor cost savings, plus time savings and convenience in both permanent plant and temporary transmission systems. When compared to copper's bandwidth and distance limitations, fiber transmissions have no practical limits for these demanding applications.

4.1.2. Broadband Infrastructures

The key elements of a broadband network include both the feeder and distribution infrastructures. While the feeder plant moves traffic between a head end or central office and a neighborhood, the distribution plant moves traffic between the feeder network and actual customer locations. Modern feeder plants being deployed increasingly use all optical media, while the distribution plant still uses copper media, generally coaxial cable or twisted pair wire. The design of a hybrid network is crucial because it sets fundamental limits on the amount and types of traffic that can be delivered over the entire network.

The serving area concept, which is most important when dealing with the hybrid plant design, represents a revolution for CATV design and yet is familiar territory for telecommunications providers. The hybrid combination of optical fiber and coaxial technologies makes it technologically feasible to employ a single platform to deliver all types of information and interactive services, ranging from 64kb/s voice up to the most bandwidth-intensive telemediation applications.

When we consider the effect of the transition from analog to digital information transmission, throughput demands, or the amount of data traveling through the network, increases dramatically. In a domain where a single frame of digital video delivers approximately 921kb of information for transport, the need for increased transmission and processing capacities becomes obvious.

In the world of telecommunications, trends in commercial transmission rates have seen a quantum leap. Commercial access to high-speed carriers has increased while available capacities continue to rise. A glut in "dark fiber" lines continues to lower access provider costs which are, in turn, passed on to the consumer. While for the last decade telecommunications providers and consumers have enjoyed an annual doubling in transmission capacities, the global implementation of optical media has increased the optical media penetration rate dramatically.

The potential capacities promised by new optical media and data communications technologies such as fiber in the loop, fiber to the curb, fiber/coax, and asynchronous transfer mode (ATM) are dramatic. Transmission capacities have leaped from 90Mbp/s in the early 1980s to as high as 100Gb/s (100 billion bits per second) over an individual single-mode fiber in 1994 experiments at NTT Laboratories, Japan. (Single mode fiber is designed to pass one frequency of light (red) and is currently used primarily for long-distance networks.)

Alternative technologies such as compressed video over copper (ADSL) and ISDN have the advantage of enabling end-to-end digital transmissions and connections while retaining internetworking capabilities with switched public telephone networks. However they are limited in their capacities and cannot support symmetrical broadband ([3]1.544Mb/s) information services. ADSL, while supporting the existing POTS infrastructures, is capable of delivering one-way services by way of a downstream maximum of 6Mb/s. Upstream telecommunications, however, is limited in bandwidth to a maximum of 384 kb/s. Broadband wireless data communication technologies such as wireless cable (DBS; MMDS, multichannel multipoint distribution services; and LMDS, local multipoint distribution services) are capable of delivering high-bandwidth services, but all interactivity must use an alternative network provided by telecommunication carriers. Multi-gigabit low elliptical orbit satellites (LEOS) hold the greatest potential bandwidth for connecting consumers by way of wireless media to fiber backbone networks. Estimates predict as much as 3 Gbit/s asymmetrical capacity will eventually be made available per carrier channel. One irresolvable limitation remains, however; latency delays in data transmissions may eliminate the technology as a viable option for real-time, high-bandwidth applications which demand nominal delays.

Ultimately, the technology leading the broadband field is optical media. When the potential for multimode fiber, the proposed standard for tomorrow's mixed voice, video and telemediation networks is considered, the potential in bandwidth availability is unimaginable. It has been shown that optical media have virtually no inherent limitation in their potential throughput capacities. Limitations lie principally within the switching and transmission technologies employed. While a technological impossibility today, the ultimate potential of optical media is simply incomprehensible. It is not difficult to imagine, for example, an individual optical media fiber capable of utilizing spectrum frequency division multiplexing, much like today's RF systems, that refract light carriers throughout the entire visible spectrum (R, O, Y, G, B, I, V). With spectral technologies in place, a single multimode fiber could produce a broadband transmission medium capable of surpassing any data transmission technology we have today. Assuming that each color carrier was multiplexed and transmitted at the experimental rates which had already been successfully achieved over single-mode fiber (100Gb/s), one strand of multimode fiber could theoretically be capable of a throughput capacity as high as 700Gb/s. A data throughput capacity of 700Gb/s is compre-hensible, even if they currently surpass the capabilities of even the most advanced networks today. When the economics of optical media are taken into account, however, we realize that to be cost efficient, fiber optic media must be bundled together (current single mode bundles consist of 244 fiber strands). The potential data capacity then for a single, fiber optic cable/bundle explodes to nearly 170 terabits or 170 trillion bits per second (244x700Gb/s). Multiple bundles such as those found in undersea installation cable plants could ultimately usher in the era of the petabit (Pbit/s = 1,125,899,906,842,624 bit/s; 17,592,186 64 kbit/s voice channels) networks.

4.1.3. Switching: ATM (The Holy Grail)

The viability of the evolving asynchronous transfer mode (ATM) switching technology standard for use in public data and telecommunications networks is increasing. ATM provides a high-bandwidth, low-delay, packet-like/cell-based switching and multiplexing technique for transporting user data over a variety of network infrastructures. As opposed to synchronous transfer mode (a method of digital transmission in which discreet signal elements are transmitted at a fixed and continuous rate), ATM can be used over multiple media plants, including copper and fiber. Most significantly, ATM supports simultaneous fixed and variable bandwidth services (including synchronous and isochronous data types such as voice and video) over the same network infrastructure, providing for efficient bandwidth on demand multimedia services.

ATM continues to receive support from the converging industries of telecommunications and computing. An entirely digital switching and transport technology, ATM switches and transport devices care little whether the information processed is audio, video, text, graphic, or any other data type, including several synchronous and asynchronous computer networking protocols such as IPX or TCP/IP. By providing end-to-end digital connectivity much as the original plans for broadband ISDN called for, ATM is the first accepted data networking technology to cross the boundaries of wide area and local area data networks (WAN/LAN). By virtue of its compatibility with the existing international SONET payload architecture, ATM transport technologies are destined to become the core transmission technology of tomorrow's advanced intelligent networks, providing session management and control of broadband telecommunications (at speeds from 1.544 Mb/s–622 Mb/s) while retaining internetworking capabilities with switched public telephone networks.

4.2. Media Compression

In the digital era, the previously complex tasks of signal processing and data transmission have become even more demanding of infrastructure resources. The convergence of computers, video, and other data formats has already offered a glimpse of how this new dynamic palette of interactive services can affect the way we communicate at home, work, or play. When a shift in underlying technological foundations is as fundamental as that from analog to digital (voice-based telephony to optical, binary, or digital interactive services that simultaneously include text, numbers, audio, video, and 3D graphics), the need for computer power processing and data transmission capacities increases dramatically. Digital video, as discussed earlier in this chapter, is a computer technology. As such, a balance between the required data rate and the throughput capabilities of the serving computer is critical to the successful delivery of real-time data representing the video signal.

Consider the demands:

- To represent a one second clip of full-motion, studio-quality raster video (4:2:2), ITV hardware and networking infrastructures would require, for full-frame, high-resolution 30 FPS video, an end-to-end digital connection with an effective data transfer rate of 251M bit/s or 31 M/bytes per second from server to receiver.
- For full-rate digital transmission of HDTV, infrastructure demands rise to 1,500 M bit/s.
- The television industry CCIR Rec. 601/SMPTE 125m standard for broadcast quality serial digital video requires throughput of around 270 M bit/s.

The question remains then as to what technology is available to the consumer that can handle interactive services with such high throughput requirements? The required bit rates for uncompressed full-quality, full-motion video are still too arduous for most commercial network infrastructures and processors and are currently impossible for existing consumer architectures. Only ATM which is now in its early stages of deployment is currently capable of

254

supporting such data rates (^{2} 622Mbps/OC48). BISDN telephone lines in the United States can handle approximately 1.544 Mbit/s and in Europe it is 2.048 M bit/s. Even the fastest commercial LANs (FDDI) from the computer networking world are capable of only 100–150 Mb/s.   A daunting gap still exists between the required data rates and the capabilities of current technology.

Advancing and converging computer and available telecommunications technologies continue to chip away at the chasm. From the collaborative efforts of the computer and telecommunications industries has come an enabling technology that promises to change the way we collect, process, and deliver our media over narrowband communications technologies and gives increasing channel capacity to broadband networking infrastructures. Compression allows any data type to be manipulated, altered, and transmitted in a more efficient manner than its original form. The codified process of reducing redundant information in a message called an algorithm. Regardless of whether the execution of the routine is embedded in software or hardware, the processing routines of the compacting scheme or algorithm is referred to as the codec (encoder/decoder).

4.2.1. A Standard Approach

Standardized video compression techniques are required for the sharing of video information between platforms. Fortunately for the consumer, the

International Standards Organization (ISO) has reached some level of success in standardizing three particular algorithms. Although their applications differ, the characteristics specific to each algorithm—JPEG, MPEG and P*64—have been outlined. Implementation of the standards remains the responsibility of the manufacturer. While manufacturers have agreed to support these standards in principle, they also continue to develop new and more efficient proprietary algorithms.

4.2.2. MPEG I & II

Developed with the ISO/IEC, MPEG I & II are intended for the distribution and playback of digital movies and prerecorded video. With consideration for data rates and capacities of personal computers, interactive television, and other low-bandwidth delivery methods, the emphasis in algorithm design was on maintaining suitable quality with maximum compression.

Characteristic of the full-motion (MPEG, P*64) codecs is the use of interframe compression, predictive coding, motion estimation, and picture interpolation. Interframe compression is the starting point for the MPEG algorithms as referenced by a key frame. Additional techniques such as motion estimation, predictive coding, and picture interpolation are also utilized to reduce unneeded bits and bytes. Essentially the combination of these techniques allows for the algorithm to capture and compress data representing movement between frames in terms of differences only, thereby reducing the required data rate. The process also allows for the nonsequential arrangement of frames and reduced image size on playback.

The MPEG I standard is currently deployed in the first generation of digital CD-ROM based games and the new video-disk technologies. The standard for an even higher resolution all digital, non-interlaced high-definition television (HDTV) is expected to become a reality when approved by the FCC in early 1995. MPEG II has been chosen to reduce the even higher throughput demands created by the new 720 line (1,280x720x24x30/ 1.4Gb/s) progressive scan (non-interlaced) consumer grade system. Following the compression formats above, a "crunched" HDTV signal can be multiplexed onto a proposed 22 Mb/s 6MHz RF signal for broadcast delivery.

4.2.3. CCITT H.261 (P*64)

Developed by the ITU-TSS (formerly the CCITT) H.261 or P*64 is an international telephone video conferencing standard. Unlike JPEG which was developed for high-quality image compression, the goal of P*64 is to maintain a fixed maximum data throughput, while minimizing encoding and decoding delays over one or more 64 kb/s telephone connections.

In order to maintain throughput, P*64 encodes real-time 30 fps video and audio at variable bandwidths and smaller than full-screen displays.

User-definable parameters allow control of the compression process and resultant bit rate to avoid filling a hypothetical data buffer.

Depending on whether the goal is high picture quality or smoothness of motion, at low connection speeds, this standard can be adjusted accordingly. Adjustable parameters affecting quality include customized block significance and quantization step size, frame dropping (reportedly when used on lines with speeds in excess of P*8 /5.12Mb/s, latency is imperceivable).

Additional compression algorithms are being developed at an ever increasing pace. The international working group that develops video conferencing standards will issue, in 1995, the first standard for interactive video over an analog phone line. The videophone standard (H.32P), as it will be referred to, integrates the H.361 video codec algorithm and is designed to foster the interpretability of a new generation of video equipment for laptop, desktop, hand-held, and mobile devices.

An even higher resolution format, the Digital Film Format (3000x1536x24x24/2.65Gb/s), is being proposed by the Technology Council of the Motion Picture-Television Industry to support electronic postproduction and maintain digital archives of Hollywood's products with sufficient resolution to support any electronic transmission format that is likely to emerge in the next century.

5. NEW MEDIA MARKETS

The use of public switched networks for highly interactive telemediation will bring forth an entirely new era of telecommunications possibilities. With the introduction of innovative and enabling technologies such as optical transmission media, ATM switching, and video compression, information technology is simultaneously pushing and redefining the boundaries, capacities, and capabilities of tomorrow's telemediation netscape. Much has been speculated about the futuristic information superhighway, I-Way, etc., and what a true telemediated netscape would look like to the uninitiated.

There exists today a low bandwidth alternative that can offer a glimpse of what a telemediated future may look like. The Internet, a global matrix of computer networks, currently connects more than twenty million independent networks or domains around the world. Behind each domain lies an unpredictable amount of subscribers with varying bandwidth capacities and demands ranging from dial-up 2400bps access to full-blown 155 Mbit/s ATM backbones.

In contrast to the futuristic visions offered by the telecommunications and cable operators, the architecture of the Internet is derived from both the data communications and computational paradigms. All services available through the Internet have been based upon the notion of equal and open access, where information is data, not analog signals, flowing from either point-to-point or point-to-multipoint destinations. In a significant departure from the telecast paradigm, and the proposed commercial offering mentioned previously, the Internet includes support for both peer-to-peer and client/server application paradigms over a single data communications infrastructure.

Utilizing an open standards approach based upon the TCP/IP protocol set , the Internet supports diverse services through integrated access arrangements and the definition of a limited set of standard, multipurpose interfaces for equipment vendors, network providers, and users. TCP (transmission control protocol) is a transport layer connection–oriented, end-to-end protocol that provides reliable, sequenced, and unduplicated delivery of data to a remote or local user. IP (Internet protocol) provides for transmitting blocks of data between hosts identified by fixed length addresses. While IP has no mechanisms to augment end-to-end data reliability, flow control or other services, these routines are provided for by other protocols. TCP/IP may be easily encapsulated into other synchronous or isochronous transport protocols such as ISDN, ATM, or Ethernet. This design closely parallels the basic architecture of public switched telephony networks. As such, it is not surprising that recent innovations have extended the Internet's data networking capacities to support both broadcast and telecommunications capacities and capabilities.

In a networked digital environment such as the Internet, each member of a community has full access to multimedia and telecommunications at all times. The Internet's World Wide Web provides this access. The World Wide Web, developed originally by scientists at the Particle Physics Research Laboratory in Switzerland, envisioned a method of providing access to nontextual-oriented information residing on disparate, incompatible databases across the Net. The result of these efforts is the World Wide Web, a meta-layer of the Internet through which interlinked information resources may be accessed.

Through the use of the HyperText Transport Protocol (HTTP) and a client-server architected "browser" containing a superset of the numerous TCP/IP protocols (Gopher, Archie, SMTP, FTP, and others), a global information system has been created. Individual items of information stored on separate and dissimilar computers around the world can be linked together through networks, automatically or by a user, as part of an information-composition process. Under the HTTP design, each individual element of information, whether a text, audio, video, or 3D animation, is seen as a single element of a massive global database, with its own unique address. These addresses are utilized then to link other pieces of information or locations around the globe into a single "virtual" document.

Access to the World Wide Web can be obtained by establishing a user account or connection to a local Internet access provider, who, in turn, connects individual users to the Internet backbone. Once on-line, users have equal, symmetrical, and simultaneous access to full-motion video and film clips, audio events, computer-generated imagery, animation, scanned images, and a variety of other media types, through which they can navigate with the point-and-click ease of the graphical interface on their home computer. At present, the majority of this information is not being provided by any one single content provider, but by millions of users around the world for the price of a local phone call to their access provider. For an additional fee, users may make available to a global community their own information.

The adoption of the WWW has fostered a personal communications revolution not dissimilar to the desktop publishing revolution of the 1980s and could have impacts as far reaching as the telephone for wide area commerce. The capacity to handle all media types and interconnect a global matrix of minds is simultaneously expanding and connecting a truly global village. Internet-based communications are fostering innovative solutions for new media applications as an alternative to traditional telecommunications service providers. By providing full integration of dynamic data for highly interactive telemediation, services such as media on demand, interactive programming, remote application control and monitoring, media/data storage/retrieval, informational services, and peer-to-peer telecommunication services are rapidly being made available to a global market.

Many of the services envisioned by the telecommunications and broadcast/cable industries are being explored over the Internet. The MBONE, an experimental media-capable backbone supporting isochronous data transfer, has made such shareware applications as Netphone, the Internet equivalent of the PictureTel, and a host of other broadband applications including on-demand/unscheduled synchronous two-way audio/video conferencing at varying resolutions possible. Services can be provided in real time, from desktop to desktop with respective sharing of tools and data. Narrowband two-way synchronous audio/data communications, point-to-point communication/file transfers, remote application control and data sharing, point-to-point and point-to-multipoint data and full-motion media telecasts, and desktop video conferencing applications and services are all available today, on a global scale, over the Internet.

It is important to note that the Internet is an international virtual network with no true regulatory body, is subject to no international legislative powers, and is easily capable of bypassing any disruptions in service. Accordingly, users with intense communication demands are already using the Internet to send computer data across the globe.

Companies are starting to use the Internet as a pseudo-private backbone to carry fax, voice, and video traffic between sites, bypassing traditional exchange carriers. New products such as IP/FaxRouters and software-based telecommunications packages allow packet-based, half-duplex telephony services while retaining the simple-to-use interface of a common fax machine. With an additional sound card, microphone, and video capture board, Internet-based telemediators currently have access to desktop video conferencing services that the telcos have attempted to provide for years.

As information is increasingly transformed into raw data, content for all industries will be available as bits and bytes. Ubiquitous access, portability, demand for interactive multimedia products and services, deregulation, and the technological convergence of computers and telecommunications have changed the roles of service subscribers and users from that of passive consumer to active shapers of market demand. Combined with the erosion of the LEC's "natural monopoly," the groundwork has been laid for a new media revolution.

6. CONCLUSION

The issues discussed in this chapter are intended to inform decision-makers in the converging worlds of telecommunications, broadcasting, and computing of the potential opportunities and conflicts that may arise from the union of such historically divergent industries and mind-sets. From the discussions and observations made in this chapter, it is clear that the information economy is generating an environment of dynamic change and innovation. Technologically enabling innovations, economic factors, divergent industry perspectives with their accompanying models of information flow, future infrastructure needs, and communication service designs of the evolving telemediation industry, are all becoming inextricably linked with each other through the convergence of information technologies.

In the information age, it is the linking of information and the end user, its raison d'être, that provides economic growth and stability. Technological developments and the digitization of all media types continue to revolutionize entire industrial processes, from the placing of a long-distance call to the delivery of a product or service to the end user. This process of supplying the market for information goods and services will continue to evolve at a faster and more efficient pace depending on the way previously divergent industries interact with one another. The continued expansion and stabilization of the information access and provider industries will depend upon their respective abilities to balance the interaction between market demand and product/resource development.

Deregulation of telecommunications initiated in the United States to meet market demand for enhanced services has resulted in a competitive and more efficient industry. Close links in the international marketplace have pushed the trend to deregulate to a global level. While the economic and global effects of the international dimensions of regulatory changes have yet to be clearly determined, one fundamental factor emerges: highly regulated monopolistic structures cannot respond quickly enough to take advantage of technological changes.

Real economic costs and benefits are also closely related to each of the respective industrial paradigms discussed here. For telecommunications access providers, technologies are advancing rapidly, outdating their aged infrastructures. Yet as the cost of information transmission continues to decline, the demand for bandwidth increases. In an era increasingly inundated with high bandwidth, telecommunications capable multimedia information processing (telemediation) applications and symmetrical communications services are what will be in demand. Dial tone service alone will not generate sufficient revenues to sustain the telecommunications industry in the next century.

Industry leaders can no longer afford to emphasize a narrowband (64k to 1.544 M bit/s) voice only or "limited experience port model" of interactivity for future global markets. However, historical precedents established in the differing industrial domains that will be offering services continue to be carried over to the new media markets and the perceived role of consumer participation remains extremely limited. With only a glimpse at the mushrooming Internet subscription rates, this oversight becomes glaringly obvious. People want the ability to communicate freely and globally in both directions, on demand, with a variety of media and computational support services.

In the end, information carriers, producers, distributors, and end users alike must come together in contributing to the success of a telemediation netscape. Technological advances have made the upgrading of existing infrastructures an economically viable and necessary reality in this century. Ultimately, however, only an awareness of multiple information infrastructures that suitably address users, resources, and strategies from previously disparate industries, coupled with alliances between individual and cross industry partners, will determine the collective effectiveness of telemediation in an era of petabit networks.

NEW MEDIA GLOSSARY

Technology is changing the way we communicate and redefining the way we speak and write. As the era of communication secedes to the new media–information age, the efficient transfer of information, spoken and written, will mark the informed and industry leaders. As telecommunications, cable-television, and network providers increasingly find themselves in the same business, there remains a significant language barrier. To assist in understanding the evolving language of the new media revolution, several key terms are listed below.

Advanced Broadcast Video Services (ABVS): A compressed video transmission service offered by the Pacific Bell RBOC. This real-time service, with a data stream of 45Mb/s allows transmission of an NTSC signal with four audio channels compressed at a 3:1 ratio. This full-motion, full-screen video arrives at the other end after a delay of about 20 milliseconds.

Asymmetric communications: Full duplex communications that transmit data in one direction at one speed and simultaneously in the other direction at another speed.

Asymmetrical compression: A compression system that requires more processing capability and time to compress data than to decompress data. Such systems are typically used for mass distribution of media where delivery and bandwidth limitations are of concern.

Asynchronous transfer mode (ATM): A multiplexing and switching technique that organizes information into fixed-length, 53-byte pieces or cells, sending it randomly through a network, where it is reassembled cell by cell at the other end. The transfer mode is asynchronous in that the recurrence of cells depends on the instantaneously required bit rate. The advantages over the standard DS-3 signal used today is that bandwidth is essentially unlimited and compression is unnecessary.

Asynchronous transmission: A signal lacking synchronization with some other reference signal. Time intervals between significant events are not equal.

Basic rate access: Access to an Integrated Services Digital Network (ISDN) that provides data transmission at rates to 144 kbit/s. The information stream is divided into two 64 kbit/s B-channels and a 1x16 kbit/s D channel. The B channels are used for voice or data. The D channel carries control and signaling information to set up or take down connections, and can also carry X.25 packet data at rates up to 9.6 kbit/s.

Broadband communications: Voice, data, and video communications at rates greater than 1.544 Mbit/s (T-1); that is, greater than wideband rates.

Broadband ISDN (BISDN): Wide bandwidth ISDN network architecture designed to deliver switched digital voice, data, and video signals into businesses and homes over a single transmission medium.

Broadband: The quality of a communications link having essentially uniform response over a given range of frequencies. A communications link is said to be broadband if there is no perceptible degradation to the signal being transported.

Client server networking: An architecture for data communications applications and interactive networks in which the client is the requesting machine (set-top box/user) and the server is the supplying machine (telco/cable head end). The client provides the user interface and performs some or most of the application processing and request brokering. The server maintains the databases and processes requests from the client to extract data (video, home shopping services, etc.) from the database.

Collaborative service model: Model for provision of services to the consumer market in which communications between data and service providers are

peer-to-peer in nature, bi-directional, symmetrical, synchronous, and in
real time.

D-1 video: The generic name used to describe a digital component video
recording system. Commonly used to refer to an international standard for
component digital television (CCIR601) that defines the sampling systems,
matrix values, and filter characteristics for both Y, Cr, Cb, and RGB
component digital television.

DCS: A digital cross connect system allowing customers to reconfigure their
network map on demand. Fundamental technology for virtual circuit
switching.

Digital hierarchy: A system of standardized transmission rates for digital
signals using time division multiplexing techniques over copper media.
The rates are: DS0, 64 kbit/s (one voice channel); DS1, 1.544 Mbit/s (24
voice channels); DS1C, 3.152 Mbit/s (48 voice channels); DS2, 6.312
Mbit/s (96 voice channels); DS3, 44.736 Mbit/s (2016 voice channels).

Integrated network access: A network architecture for special services that
extends a digital carrier into a distribution loop to eliminate the need for an
analog connection between distribution and interoffice facilities. The

carrier terminates either at a remote terminal or at the customer's premises.

Interactive multimedia: A term that refers to the delivery of multimedia information in a nonlinear format, where control is vested in the end user. Varying levels of user control can be classified along a continuum of applications from passive to interactive to adaptive.

Integrated Services Digital Network (ISDN): (1) A series of standards for a digital local loop network to transmit voice, data, and low resolution, nonreal-time video (64 kb/s) over phone lines to home or office; (2) a network architecture that enables end-to-end digital transmission and connection. The network supports diverse services through integrated access arrangements and defines a limited set of standardized, multipurpose interfaces for equipment vendors, network providers, and customers. Internetworking with a public switched telephone network is retained. Access is provided at two levels: basic rate (BRI=144 kbit/s) and primary rate (PRI=1.536 Mbit/s).

Isochronous transmission: Time dependent. Real-time voice, video, and telemetry are examples of isochronous data. A signal in which the time interval separating any two instants is theoretically equal to the unit interval or to an integral multiple of the unit interval.

Limited experience port service model: Model for provision of services to the
consumer market in which communications between data and service
providers are client-server in nature, bi-directional, asymmetrical, and
asynchronous.

Multicast: A feature of DCS that allows one incoming signal to be replicated
and sent to many destinations.

Multimedia: A term that refers to the delivery of information that combines
different content formats in a single digital domain, including motion video,
audio, still imagery, graphics, animation, text, and data.

Multiplexing: A hardware/software process to bundle a number of lower-
speed channels for transmission over a higher-speed system. A network's
backbone transmission system will often multiplex a number of DS-3
signals into channels of up to 2.4 Gb/s (OC-48).

Narrowband channel: A transmission channel whose bandwidth can be
wholly contained within a 4 kHz voice channel.

Narrowband: A communications channel of restricted bandwidth, often
resulting in degradation of the transmitted signal.

Optical Carrier Level 1 (OC-1): In the synchronous optical network (SONET), the optical signal at 51.840 Mbit/s that results from the conversion of an electrical Synchronous Transport Signal Level 1 signal.

OC-12: 622 Mb/s optical channel capable of carrying uncompressed D-1 Video signals (270 Mbit/s).

OC-3: The optical carrier 3, optical channel interface to SONET, operating at a data rate of 155 Mb/s. Data rates capable of transporting transport D2 video (145 Mbit/s) uncompressed.

OC-48: An optical channel backbone that can carry a number of channels (up to 2.4 Gb/s) in individual channels of 45Mb/s.

Optical channels: Optical interfaces to SONET telecommunications standard. Connect rates range from OC-1 to OC 48.

Peer-to-peer communications: Communications in which both sides have equal access to and responsibility for initiating a symmetrical session.

Petabit: See Space/time.

Primary rate interface: In the Integrated Services Digital network (ISDN), a channel that provides digital transmission capacity of up to 1.536 Mbit/s (1.984 Mbit/s in Europe) in each direction. The interface supports combinations of one 64-kbit/s D channel and several 64-kbit/s B channels or H channel combinations (H0, 384 kbit/s; H10, 1.536 Mbit/s).

SONET: Synchronous Optical Network.

Space/time: The following units of measure are used to define storage and transmission capacities:

K	(kilo)	thousand	1,024
M	(mega)	million	1,048,576
G	(giga)	billion	1,073,741,824
T	(tera)	trillion	1,099,511,627,776
P	(peta)	quadrillion	1,125,899,906,842,624

Symmetric communications: Full duplex communications that transmit data simultaneously in both directions at the same speeds.

Synchronous Optical Network (SONET): A transmission standard operating at higher rates than current systems for transporting a wide range of digital telecommunications services over optical fiber. SONET is characterized by standard line rates, optical interfaces, and signal formats.

Synchronous transmission: Transmission of data in which both sending and receiving stations are synchronized. An operation that occurs at intervals directly related to a clock period. Codes are sent from the transmitting station to the receiving station to establish the synchronization, and data is then transmitted in continuous streams.

Synchronous transport signal level 1: In the Synchronous Optical Network (SONET), the basic electrical signal with a rate of 51.840 Mbit/s.

Synchronous transport signal level N: In the Synchronous Optical Network (SONET), the electrical signal obtained by byte interleaving Synchronous Transport Signal Level 1 signals together. The rate is N times a rate of 51.840 Mbit/s. Levels include STS-1 through STS 48 more commonly referred to incorrectly as OC1-OC48

T-1: Equivalent of a DS-1 channel or 24 DS-0 channels. Used heavily for corporate data communications and may be used for transmitting lower-speed, low-resolution video for noncritical video conferencing.

T-3: 28 T-1 Channels.

Telecommunications Capable of Multimedia Information Processing (Telemediation): A term used to represent the evolution of telecommunications and computer-mediated communications to a level of integrated open, communications.

Telemediation netscape: A social and physical networked environment of individuals and telemediation devices facilitating communications across national, industrial, organizational, and time boundaries.

Telecommunications Capable of Multimedia Information Processor (Telemediator): A term intended to replace the outdated single function connotation of "computers." The term is intended to emphasize the confluence of human communicative processes, multimedia technological capabilities, and advances in broadband telecommunications.

Wideband communications: Voice, data, and video communications at digital rates of 64 kbit/s to 1.544 Mbit/s.

Wideband: The passing or processing of a wide range of frequencies. Meaning varies with context. (Audio [3] 20 kHz; Video [3]6 MHz; Data [3]1.544 Mbit/s).

277

REFERENCES

Apple Computer, 1994, *Multimedia Demystified*, Random House Electronic Publishing, New York.

Arnst, C., Kevin Kelly, and Peter Burrows, 1995, "Special Report on Telecommunications: Phone Frenzy", *Business Week*, February 20, pp. 92–97.

Birkmaier, C., 1993, "Deeply Disturbing", *Videography*, December, p. 36.

Cole, A., 1993, "Final HDTV System Comes Into Focus", *TV Technology*, December, p. 1.

Doyle, B., 1994a, "Crunch Time for Digital Video", *New Media*, March, p. 47.

———, 1994b, "How Codecs Work", *New Media*, March, p. 52.

Freedman, A., 1993, *The Computer Glossary*, 6th ed., Computer Language Company, New York.

Gaggioni, H., 1994, *Video Compression: A Primer on Bit Rate Reduction Technology for Video Signals*, Primer No. Sony Corporation of America, New York.

Harasim, L. M., (ed.), 1993, *Global Networks*, MIT Press, Cambridge.

Hazell, J., 1993, "Fiber Optics and the Splice of Life", *TV Technology*, December, p. 21.

International Radio Consultative Committee, 1990, *Encoding Parameters for Digital Television Studios*, No. CCIR Recommendation 601-2-90, International Radio Consultative Committee, Geneva.

Jussawalla, M., 1992, *The Economics of Intellectual Property in a World Without Frontiers: A Study of Computer Software*, Greenwood Press, Westport.

Karpinsky, R., 1994, "Fiber in the Loop Architecture", *Telephony*, February, pp. 28–30.

Kessler, G. C., 1990, *ISDN: Concepts Facilities and Services*, McGraw Hill, New York.

Littlejohn, S. W., 1991, *Theories of Human Communication*, Wadsworth, Belmont.

Locke, J. G., 1994, *Digital Video in the Era of Petabit Networks*, unpublished paper, University of Hawaii at Manoa, Honolulu, Hawaii.

Pinkham, R., 1994, "Combining Apples and Oranges: The Modern Fiber/Coax Network", *Telephony*, pp. 32–36.

Quinell, Richard A., 1993, "Image Compression, Part I, II, III", in *EDN: Electronic Technology for Engineers and Engineering Managers Worldwide*, vol. 38, no. 2, pp. 62–71.

Society of Motion Picture and Television Engineers, 1992, *Component Video Signal 4:2:2–Bit Parallel Digital Interface*, ANSI/SMPTE 125-M 1992 Standard for Television, Society of Motion Picture and Television Engineers, White Plains, New York.

Whitaker, J., and Harold Winard, 1992, *The Information Age Dictionary*, 1st ed., Intertec Publishing Corporation and Bellcore, Overland Park, Kansas.

SATELLITES BID FOR THE GII

MEHEROO JUSSAWALLA

1.   INTRODUCTION

The fast pace of change in information technology has regulators the
world over racing to catch up. Never before have so many changes taken
place in the regulatory regimes of sovereign states and the international
system, be it in the structure of the industry itself or in the allocation of
resources. The United States has been leading the "convergence revolution"
and enticing users into cyberspace with information systems like the Internet
and World Wide Web. On the one hand, the magnetic frequency spectrum is
being allocated either through auctions or through allocations and on the
other, there is a scramble for the geostationary orbit regardless of regulations.
The fact remains that profits loom large in the coming age of convergent
technology based on digitization. Already information is being moved across
continents in real time and at the speed of light. The electronic highways that
are encircling the globe are actively using the convergent compression
technologies of multimedia that are rendering every conduit of telecommunications interactive. These services are being offered in a multitude of
operations allowing CD ROMS, virtual reality, and camera-based and

voice recognition computers to add to the excitement of speeding on the information superhighway. In this competition for global markets, the submarine fiberoptic cable industry and the television cable industry are challenging the supremacy of satellites. Telecommunications market analysts are predicting that wireless communications will be the largest growth area at the turn of the century. With the digitization of mobile communications, broadcasters as well as telecommunications operators are invading the data transmission market creating a $10 billion industry for data alone in the year 2000. This is part of the worldwide market for telecommunications equipment and services estimated at $3 trillion of which the wireless portion is predicted to be $600 billion. Imagine the choices for consumers when broadband conduits will integrate wireless voice, data, and compressed video images and deliver them to offices and homes using PCS (personal communications system). Despite the challenge that is emerging from DBS (direct broadcasting satellites), the cable industry is taking advantage of the convergent technology to explore new ways of delivering their services in the interactive mode. Yet both satellites and cable have survived in a symbiotic relationship over the years which has made them complementary rather than competitive. Even as satellites through their DBS and DTH (direct to home) services are breaking the monopoly of cable in providing over one hundred channels of TV viewing, they are helping the cable industry by letting it provide low-cost return paths for two-way services. Jointly, both these rivals stimulate the multimedia marketplace.

2. SPECIFIC BENEFITS OF SATELLITE USAGE

Quite apart from the economic benefits emerging from the use of satellite technology such as its cost-insensitivity to distance, its ability to communicate over hard-to-reach terrain and to integrate islands that form part of difficult geopolitical configurations, there are newer benefits arising from new technologies on board satellites. For more than a decade, governments and business corporations have been profiting from the use of VSATs (very small aperture terminals) which are portable and ideal for data transmission. For example, the People's Liberation Army in China is manufacturing VSATs and locating them over urban and rural areas of the country.

Now we have "pizza-sized" satellite dishes (19 x 20 inches) that allow the user to unplug from the cable and receive digital signals directly from the satellite. In the United States alone, 590,000 dishes were sold in 1994 (Flannagan 1995). DBS, a joint venture of Hughes with Thomson Consumer Electronics and DIRECTV, are used to broadcast to multiple sites such as retail stores, restaurants, and banks (including their branches). The limitation is that DBS dishes are receive-only, whereas VSATs are two-way. Even personal information systems are being tested for orbital linkage via DBS which is cheaper than and bypasses land lines.

In 1993, DIRECTV launched one of two satellites capable of beaming 150 channels direct to American homes, from an orbit of 22,300 miles above the equator. European and Asian satellites have also adapted their systems to DTH technology. The receiving equipment for DBS is not only smaller and cheaper, but because of the digitization of the satellite, it provides CD-quality sound and pin-sharp pictures. The cost of the satellite is $600 million and currently the dish costs $700 inclusive of the set-top box to unscramble the signals. Already 700,000 subscribers have signed up in rural parts of the United States where there is no cable TV. Seeing the success of DBS, large cable companies like Telecommunications Inc., the world's largest cable operator, are ordering their own DBS birds to distribute programming.

While in Europe there is too much competition in satellite manufacturing with Aerospatial incurring losses, the market is booming in the Asia-Pacific. Greater scope for DBS is being generated by the multimedia convergence. In a 1994 study, *DBS: The Time Is Now*, the National Association of Broadcasters noted that the Asian market is forecast at 2.5 billion subscribers with primary growth markets in China and Taiwan.

One significant trend in satellite systems is the desire to enhance competition not only through separate systems but by privatizing Intelsat. The problem is, can a cooperative of over 121 countries be privatized and if so, can its services be delivered to its member countries at the same cost and

with the same quality as done over the years by Intelsat? Competitors of
Intelsat feel that there has been an unfair competitive advantage given to
Intelsat. Its market dominance has been challenged and its restructuring is
called for. Intelsat is also facing competition from the mobile satellite
systems. But Intelsat plans to provide newer services on eight new satellites
to be launched over the Pacific Ocean region alone. Its three existing
satellites are located at 174, 177, and 66 degrees east, and they are capturing
a large share of the VSAT market.

In large part, the debate over privatizing Intelsat and Inmarsat
(International Maritime Satellite) has been caused by increased competition,
the elimination of monopolies, and new technologies that are rapidly changing
the satellite industry. By 1996, there will be seven separate system satellites
competing with Intelsat according to Crockett (1995), President of COMSAT.
The competition is also intensifying from the submarine fiberoptic systems
that are linking continents from Europe and the Middle East to Asia and
Oceania. At the same time that the marketplace rules and portable earth
stations have replaced national telecommunication borders, both Intelsat and
Inmarsat are saddled with top-heavy bureaucratic structures that need to be
dismantled to make them more flexible in shifting policies. While it is true
that countries around the world are benefiting from the rate reductions that
have taken place in satellite services provided by both these organizations,
the economic and political climate within their members have also changed.

It may be possible to change their structure into multinationals in which
decisions made will be based on business requirements rather than on each
country's investment. In any case, over the next few years the most
important trend in the telecommunications industry will be the privatization
of Intelsat.

3. SATELLITE MANUFACTURING ISSUES

As communication services are becoming more technically driven than
before, deregulation and diversification are complicating the industry, and
strategic alliances between manufacturers, financing organizations, and
system operators are becoming more common. Manufacturing costs are
being reduced and just-in-time delivery schedules are being more closely
followed. Better services are being offered by manufacturers to the clients in
the insurance and finance sectors. There was a time when there were just a
few manufacturers in the United States with Hughes being the pioneer along
with Ford Aerospace (now Space Systems Loral) and later, Martin Marietta.
Canada and Russia, along with Europe, manufactured satellites. Today
however, Japan, China, and India are manufacturing their own satellites for
remote sensing and communications. Many developing countries
manufacture their own receiving equipment as well. While the market for
satellite transponders has grown at an exponential rate, so have the suppliers
depending on the success of the launch vehicles. While NASA dominated the

market for commercial satellite launches, it is now being replaced by competitors from the private sector in the United States and by Arianspace of France, Long March of China, and the Proton from Russia.

The demand for satellites has boomed because of mobile satellite systems being planned around the globe and by the little and medium low earth orbiting satellites (LEOS). In Europe, the Martin Marietta Astro Space system is facing competition along with the fact that many clients have limited experience and need help from the manufacturer for their launch-related requirements. It appears as if there is an overcapacity in satellite manufacturing in Europe, and Aerospatiale has escalated this competition into a crisis. It is discussing mergers with other companies in the aerospace industry, such as Matra Marconi Space.

The future of the communications satellite industry rests on the growth of digital satellite systems which are the fastest-growing consumer electronic device in the evolution of the industry. Competition for this part of the satellite industry is emerging from the distribution of video in broadband, two-way, all-digital interactive systems being built by telcos and cable companies.

There is considerable inequality in the manufacture of satellites and the building of launch systems. For example, the United States holds 70 percent of the world market while launch equipment builders only hold a 20 percent share. This affects the ability of the satellite manufacturers to sell spacecraft. The U.S. government limits the number of low-cost launches from China and Russia which can be used by American satellites. Martin Marietta builds the Titan and Atlas rockets. Delta rockets are also built in the United States. If launch prices are arbitrarily kept high by the United States, it prevents the proliferation of affordable satellite programs around the world. At present, the cheapest launches are provided by the Russian Proton, the Chinese Long March, and Lockheed's new launch vehicle.

4.   MOBILE SATELLITE SERVICES

However, mobile satellite services (MSS) are widely affirmed as the most effective way of enabling people to communicate over widely dispersed areas. Successful mobile systems such as Inmarsat and Optus Mobilesat are reaching regions that have never been served before. By the end of 1995, new MSS services will be offered in America by the Mobile Satellite Corporation and in Canada, by TMI Communication. Likewise, the Secretary of Communications in Mexico will commercially operate an MSS system for the country's business users. For a seamless satellite network for MSS, compatibility of equipment will be essential. This is what Comsat Mobile Communications is aiming for in enhancing its revenues from satellite-delivered, handheld, mobile communications services known as Inmarsat P.

These will provide reliability and widespread coverage and thereby fill the gap in rural and remote areas for dial tone. Just as for Intelsat, the policy issue is of privatization, even though its charter has been fulfilled by its services to the low-income regions of the world. In Latin America, there are sixty to seventy million households that are being linked by DTH and mobile systems. Mexico's Solidaridad L-band satellite is offering mobile services. Similarly, Argentina's Nahuelsat system will become operational in 1997 offering both space segments and mobile services.

In the forefront of mobile systems are the plans for the Big LEOS, most of which are being launched by U.S. firms. Satellites play an important role through LEOS in the wireless revolution to cover rural areas at lower costs than land-based systems. Ventures like Iridium and Teledesic are planning to make a big dent in the information services market. However the capital costs are phenomenal with Iridium at $3.5 billion and Teledesic at $9 billion. Even Inmarsat-P's handheld telephone system using satellites will cost $1.4 billion in the initial phase. Globalstar is another large company promising to launch an MSS in collaboration with Loral, Qualcom, Ellipsat, and American Mobile Satellite Corporation. Of all these aspirants, only Iridium and Inmarsat P have completed their first round of funding. Meanwhile Odyssey, a joint venture between Teleglobe Canada and TRW (a U.S. aerospace manufacturer), won a patent from the U.S. Patent Office for inventing the medium earth orbit (MEOS) for launching mobile satellites. This means that

LEOS will now be competing with MEOS for locating mobile satellites for portable communication systems. The difference between the two is that while LEOS will be placed at 10,000 km above the earth's surface, the MEOS will be at 20,000 km above the earth's surface. Inmarsat-P and Odyssey both claim that the MEOS will provide a better, low-cost solution for roaming handheld communications.

Currently the market for LEOS is better assured as Globalstar has offered to place twenty-eight satellites over rural China. This network of satellites over the remote areas of northern China and Mongolia are being offered at $1 a minute compared to Iridium's $3 a minute. Iridium has contracted Canada's Telesat Corporation to build three earth stations at a cost of $29 million to be located at Yellowknife and Iqualuit in North West Canada and Hawaii. These stations will monitor and maintain the orbits of the sixty-six Iridium satellites. Iridium has already obtained equity participation from the Great Wall Corporation of China and a 5 percent share from Industrial Development Bank of India. It also has investors from Proton in Russia, Nippon from Japan, STET from Italy, and from the governments of Saudi Arabia and Venezuela. Its rival, Globalstar, has its satellites operating on the newest technology of Code Division Multiple Access (CDMA) and has attracted investors like Loral, Qualcom, Pacific Telesis, and Alcatel of France; Dacom and Hyundai of Korea; Vodaphone of the United Kingdom; and Deutsche Aerospace of Germany. Smaller LEOS constellations are Elippso

and Odyssey whose applications for spectrum have been denied by the FCC
(this explains why Odyssey merged with TRW). It is hard to decipher what
the market demand will be worldwide for so many LEOS systems, but
competition will help to reduce usage charges and benefit low-income
countries.

The Teledesic system promises to revolutionize satellite services by
providing a global broadband network. If it is to provide services for
multimedia, it will have to be compatible with the fiber terrestrial network.
LEOS are better suited to supply such services orbiting at 435 miles above
the planet's surface than are geosynchronous satellites at 22,300 miles above
the equator as seamless compatibility is greater at a lower orbit. Teledesic is
committed to deploying a thousand LEOS that will use ATM (asynchronous
transfer mode) technology to move data around at 1.2 gigabits per second.
These satellites will be located so as to divide the earth's surface into 20,000
supercells with each LEOS being responsible for sixty-four supercells within
its footprint.

5. PROBLEMS OF ORBITAL ACCESS

The United Nations Outer Space Treaty declared the geostationary orbit
(Clarke orbit) "the common heritage of all mankind." This means that the
geostationary orbit was declared a common property resource in which

monopoly ownership of any form would be illegal. Yet the technologically advanced countries like the United States and Russia began to exploit the orbital resource on a "first come–first served" basis. The geostationary orbit is neither priced nor marketable, but it is finite. Unlike mineral resources, it does not deplete with use because every communication satellite located in this orbit has a life span of seven to ten years after which it moves out. This means that the economic value of the resource increases with use (Jussawalla and Tehranian 1993). As a common property resource, it becomes possible in economic terms to either overinvest in it or let it go to waste.

The allocation of orbital spaces becomes burdened with political and economic considerations. On the one hand, the ITU regulates entry to this unregulated resource, and, on the other hand, common user groups like Intelsat, Arabsat, and Eutelsat (European Telecommunication Satellite) have no legal say within the international regime even though they are the largest users of the orbital arc. The economic and social benefits accruing from the use of satellite technology are so overwhelming that developing economies have become cognizant of these values and are making huge investments in the race for space. This has resulted in a contentious battle over its ownership and use, a battle similar to the one concerning territorial sovereignty over coastal waters.

In the real world, it is impossible to create a spectrum market because of the physical properties of signal beaming. Centralized control over signal distribution is required to prevent congestion and jamming. The orbital arc is also a joint-supply resource because the magnetic frequency spectrum through which the signals must pass to become usable comes along in a package with the orbital allocation. These are not interchangeable. As a result, if market efficiency is to be attained, there must be competition in all relevant markets. While property rights are necessary for a free market to operate with optimal efficiency, country governments are led to stockpile orbits that they may or may not use.

In the 1960s, demand was concentrated over the Atlantic Ocean region; but as trade and development grew, there ensued a scramble for the orbit over the Pacific Ocean. Intelsat was the earliest consortium of countries to provide services over all ocean regions. Its vantage position in space technology enabled it to capture the most favorable locations like the mid-Pacific, the mid-Atlantic, and the mid-Indian Ocean regions. Now due to the policy of "separate systems," Intelsat and Inmarsat are to be privatized.

The race for orbital space over the Pacific Ocean became very intense when many governments started to privatize or liberalize their telecommunications monopolies in order to reap the benefits of new technologies. Satellite systems were no exception. The Indonesian government commercialized its Palapa B2R system and moved it in orbit to 134 degrees east, calling it the Palapa Pacifik Nusantra. This position was challenged by Tongasat which claimed that the orbit had been allocated to the kingdom of Tonga by the ITU. The matter went into arbitration and the Palapa satellite had to be moved to accommodate the Rimsat satellite which belonged to Tongasat. So far, Tonga has earned almost $20 million from leasing its orbital slots even though such trafficking in orbital slots is illegal (Jussawalla 1994).

Meanwhile, in 1990, the Asiasat satellite was launched and placed in orbit at 100 degrees east. It carried the popular Star TV broadcasts on its northern and southern beams. But soon after it was launched, its orbital position was challenged by Thaisat, which was licensed by Thailand to a private company, Shinawatra, to provide domestic communications and television. The dispute was amicably settled and Thaisat was located in an orbit nearby. This episode, however, is indicative of the crowding of the geostationary orbit. A similar situation occurred when another consortium of owners from China, Hong Kong, and Thailand launched the Apstar satellite on the Chinese launch rocket called the Long March. It was placed in orbit at 134 degrees east which was within one degree of both Rimsat and the Japanese satellite owned by the Nippon Telephone and Telecommunications company. Both parties appealed to the ITU and the Apstar had to be moved. As more countries like Korea and Malaysia launch their satellites and as

Panamsat also enters the orbit over the Pacific Ocean, there will be greater crowding in space.

Countries are wary of releasing their sovereignty over orbital slots once they have been obtained. As a consequence, warehousing of spectrum is causing legal disputes hitherto unknown. This affects the affordability of services for users and the revenues for the vendors. There is not enough evidence to show whether this capacity is economically justifiable in terms of demand. It is a multibillion dollar business without stakeholder rights in the common property resource. The problem for the ITU is becoming more controversial and as technology advances, the fact remains that satellite technology will become more coveted even in the age of multimedia (Jussawalla 1995).

6.   FUTURE BID FOR THE GII

There appears to be no doubt that satellites will retain their position among the dominant technologies that are aiming to build the Global Information Infrastructure (GII) and to link users both by ground-based and sky-based systems whether they are cellular mobile or geostationary. The year 1995 is predicted to be the banner year for civilian geostationary satellites. Arianspace, for example, has listed twelve launches for the year 1995. And up to 1999, seventy-one commercial geostationary satellites will be waiting either to be launched or to be constructed under contracts. This includes nine U.S., nine European, and six Japanese satellites along with twenty-three for international systems and twenty-three for other countries (Chenard 1995). Russia too will launch twenty to twenty-eight satellites over the same period.

These numbers for future launches of satellites includes mobile satellite systems for cellular mobile networks such as MSat 1 and MSat 2 and Inmarsat's 3F1 and 3F4. Lockheed will be building LEOS for Iridium and Loral for Globalstar. The Asia-Pacific region is scheduled to have two domestic satellites, Measat for Malaysia and Koreasat. U.S. West will invest $1.4 billion in the Measat venture in collaboration with Benariang, the company owning the system. Deutsche Telekom will own a 25 percent share in the new Indonesian Satelindo venture. Indonesia's high-power, commercial satellite will be the Palapa C1, and Japan's JCSat 3 will provide regional coverage. Capacity on India's Insat C satellite is being leased by Intelsat. Rimsat will use Tongasat's locations for two Russian-made express satellites over the Pacific Ocean. Deutsche Aerospace will jointly build satellites for the People's Republic of China with the Chinese Academy of Space Technology. In the Middle East, Israel is launching the Amos 1 built by the Israel Aircraft Industries. Panamsat will also launch the PAS 4 for the Middle East and Africa, both of which have been neglected markets.

In the United States, DIRECTV Inc.'s Direct Broadcast Satellite 3 will be launched in 1995 along with its competitor, the Echostar 1. Hughes will launch the Galaxy 3R to cover Latin America and AT&T will launch Telstar 402R with trepidation, as it sustained a loss when Telstar 402 was lost in orbit. The first experiment of Orbcom's little LEOS constellation will get off when the Pegasus launch rockets will put up nine satellites on each of its two booster rockets. Europe's satellites, Astra 1E, and France's Telecom 2C will also be launched before the end of 1995.

As the satellite industry speeds up the information skyway, more LEOS operators will obtain licenses from the FCC to launch their constellations. As more direct broadcast satellites are being launched by the affluent countries, the developing nations are becoming apprehensive that their orbital slots will be preempted. Until the turn of the century, the value of this market is estimated at $12 to 20 billion after which a slowdown is anticipated because of the excess capacity of transponders in the Asia-Pacific region.

Satellites are placed in different orbits and relay signals in an ascending order from L band, S band, C band, and Ku and Ka bands depending on how many transponders or channels operate in each of these bands. The slowest growth will probably be in the United States where the industry has already matured, although all private international systems are manufactured by U.S. companies and a few by Russia. The annual projected growth rate to 1998 is approximately 9 percent. As the capacity of submarine fiberoptic cables grows, there will be a gradual decline in the demand for transponders on satellites, but this may not be a lasting trend as new technologies develop for fixed and mobile systems.

To conclude, it appears as if the movement to implement a global superhighway for information flows has created an interdependence between satellites and cables that has led to the growth of multimedia usage and its all-pervasive impact on emerging economies. Policymakers must come to grips with the global markets which will impinge on their territorial sovereignty. The competition for obtaining real-time information through digital compression and converging technologies will heat up as each country tries to export its products on world markets and as scientists and academics exchange the results of their research for the improvement of human lifestyles and the upliftment of humanity as a whole.

REFERENCES

Chenard, Stephanie, 1995, "The Banner Year Ahead", *Via Satellite*, January, pp. 24–32.

Crockett, Bruce, 1995, "Privatizing Intelsat and Inmarsat", *Pacific Telecommunications Review*, vol. 16, no. 2, pp. 10–14 .

Flannagan, Patrick, 1995, "The Ten Hottest Technologies in Telecom: A Market Research Perspective", *Telecommunications*, May, pp. 32–33 .

Jussawalla, Meheroo, 1994, "Who Owns The Orbit?", *Asian Communications*, November, pp. 27–29.

———, 1995, "Speeding Through Cyberspace", *Asia Pacific Satellite*, June, pp. 30–32.

Jussawalla, Meheroo, and John Tehranian, 1993, "The Economics of Delayed Access", *Telecommunications Policy*, vol. 17, no. 7 (September/October), pp. 517–28.

TOWARD A COMPREHENSIVE POLICY FOCUS FOR NETWORK ECONOMIC ACTIVITY

MARK HUKILL

1.   NETWORKS AND POLICIES

Developments in information technology and information systems have no doubt propelled many new economic activities in the past decade as much as they have provided a catalyst to continued economic efficiency and contributed to various forms of increased productivity. Information technology (IT) and information systems (IS) applications have also entered the domains of culture and the arts and may even contribute in some cases to the expansion of political participation. But IT and IS development, similar to telecommunications and communication network development, in general, is globally a very uneven affair and is likely to remain so for many years to come. It will perhaps even be exacerbated by rapid increases in the technological prowess of developed countries with the consequence (unintended if no less real) of widening rather than narrowing the chronic technological gaps of many developing nations.

To this context we now add the possibility of ubiquitous interactive, broadband, digital connectivity between computers and various other devices hyped as the information superhighways and information infrastructures. Despite the public relations exercise in techno-optimism (a trap into which too many of us fall), the reality of such developments will necessarily increase the gaps of communication technology availability and accessibility as this added functionality to global communications will be primarily centered around the already developed economies. No single policy, even of largesse, will readily rectify this situation. Nonetheless, information highways are an ever-growing (and only recently salient) dimension to the communication capacity of human activity. As such, they have a significant potential for enormous direct and indirect economic fallout, both beneficial and detrimental, despite their relatively concentrated buildup.

If we are to establish positive, workable policies toward the building and integration of high-capacity electronic links as an essential component of the worldwide communication infrastructure, then we must look seriously at how the changes may actually come about. This means that policy which is formulated with how economic activity arises in the network environment must put a clear focus on what is different about such activities that our current systems and policies do not address. We must also be prepared to deal with a number of different policy solutions and implementations for different parts of the world. To the extent that these different policies can be "harmonized" will determine to a large extent the ability of information highways to provide a genuinely useful global communication environment to the benefit of more people than they leave out. Certainly, a "universal service" policy can only make sense globally in such a context.

As we have yet to provide a real example of an information highway, policy is understandably underdeveloped. In short, we primarily only provide policy solutions and discussion toward policy evolution based on the realities of the networks already in place. Here briefly are some of the characteristics of network systems that we currently have which resemble only in vague parts the general concept of information highways. The fractious nature of the variable components of each network goes a long way toward explaining why current policies are demarcated and narrowly focused around essentially technical issues of each network structure.

A.   Private data networks (local and wide area)

limited user groups (closed access)

very limited access

relatively high costs

usually very secure

very high reliability

internally not very private

moderately difficult to use

almost exclusively for business or government use

(within a specific sector or organization)

fairly high level of transaction activity in addition to

information exchange

very low if any interlinkage to other networks

B.   "Public" switched telephone networks (including fixed and

mobile)

highly concentrated universal service in developed

economies

very limited penetration on a world population basis

business is the heavy user; residential users more

numerous but may provide less revenue traffic

relatively low user costs (qualified)

limited digital capacity to date

moderate privacy

moderate security

moderately high reliability

very easy to use

used mainly for information exchange

transactions conducted to the extent that oral information,

fax, EDI, etc., is trusted and perceived to be private

and confidential

C.   <u>Internet</u> (largely connecting various types of computers via a common communication protocol through leased capacity of the PSTN)

> innumerable and fractious users and user groups, but still a relatively minor proportion of the world population
>
> large government and academic component which has set the "tone" and some technical rules
>
> limited availability
>
> very limited accessibility
>
> relatively high costs (qualified)
>
> especially difficult to use
>
> limited scope of business activity to date (mostly
>> establishing a presence on WWW):
>>
>> very low privacy
>>
>> very low security
>>
>> highly interactive, although not very reliable
>>
>> information exchange as a primary activity

We also have various other major communication networks, not the least of which are the broadcast networks including over-the-air terrestrial, satellite, cable, and even trials of video delivery by phone line. These systems remain a tangled mix of analog and digital forms and one-way in message delivery. In many developing parts of the world, these networks are still in a growing stage. They are all, however, structured as relatively efficient entertainment delivery systems and, as such, offer little potential for interactivity. Policies which govern broadcasting, especially ownership and control, will however come under increasing pressure to change with the development of information highways and their potential ability to also deliver interactive entertainment content.

Questions of policy formulation and implementation will continue to evolve in line with how all of these communication network systems develop and even merge to some extent. More important, however, will be to ask the critical questions of the differences between the information highway of the future compared to the networks we have now? What will define economic activity related to these differences? Who will be the major users? For what purpose? Who will be significant minor users? How will policy need to be shaped to take into account this diversity? Who will be left out? What, if anything, in terms of policy, should be done about it? All of this is somewhat exacerbated to the extent that each of our major communication network systems to date are also regulated in different ways. The largely unregulated domain of computer networking tends to engender special cost-of-use economics. This contrasts sharply with the market pricing of telecommunication and telephony services in some countries and highly controlled, public service price regulation in others. Current telecommunications also includes the policy of "settlements" accounting,

which is an important yet largely artificial international connection pricing
mechanism in favor of inbound traffic. And the broadcast networks provide
yet another level of regulatory control predicated primarily on ownership
restrictions and service licensing in a geographic area.

The question of policy development also must grapple with the duality
of the relationship between communication technology and economic activity
itself. The diversity of policy viewpoints often derives from where on the
spectrum of this relationship one is situated. That is, we may choose to view
economic activity as a driver of technology development, or, conversely, we
may see technology implementations as fostering economic activity. Both
viewpoints, and the large gray area in between, can lead to very different
policy formulations. In essence, how we view the interaction of forces of
change such as technological developments, regulation, and competition will,
in fact, narrow the scope of the policy debate in any particular locality, but
will necessarily complicate the development of policy on a regional and
international level.

On yet another level, policy will also be shaped to some extent by
overarching ethical values and the degree to which individuals, groups, and
societies stake a claim to positions related to privacy, property, equality, and
cooperation. Guiding principles such as free and informed consent,
confidentiality, as well as equitable access, will differentiate various

implementations of economic activity on information highways. These values
and principles will necessarily be different from country to country and group
to group especially where differences in culture and religion are prominent.
The extent to which these policies can also be harmonized—that is, a
reasonable coexistence tolerated—will to a large degree define our
international relations in the network environment.

Policy questions are also conditioned to a certain extent over our
perception of answers to the following types of questions: When will there be
information highways? How will we know? For whom? How will these
"whoms" create, conduct, and sustain economic activity on information
highways? How will they likely benefit? Can others benefit? If so, how?
Why will many others not benefit?

2.   CURRENT POLICY ISSUES

Given that many of the questions above have been left unanswered,
even avoided, it is not difficult to understand that current policy discussions
focus themselves largely around "comfortable" issues to which a certain
political correctness as well as available technical solutions are perceived.
These include, but are not limited to, policies which tend to be geared toward:
(1) universal service as defined by availability through industry structures
and regulation: privacy, data confidentiality, security, encryption, and anti-

hacker measures; and (2) technical standards: connectivity, reliability, channel space and frequency allocation, and switching.

These policy issues and others related to them largely ignore the major thrust of activities which (potentially) are different on information highways as compared to the various communication networks we currently have. Serious attention will need to be focused on anticipating new and different economic activity on information highways. Such a policy focus will no doubt be rife with a large measure of uncertainty revolving around the question: will (not should) business lead in the development and/or use of information highways? This may be likely in some developed countries, but is highly unlikely elsewhere. If they do, then policy will necessarily need to be geared toward more substantive issues of business conduct and regulation in the network environment, rather than current policies which are aimed largely and superficially at technical network development.

There may be at least one certainty, however, that will act to catalyze a new policy focus. That is, the type of economic activity which eventually emerges on information highways will itself begin to drive the need to engage different policy issues from those upon which we are currently focused.

3.   ECONOMIC ACTIVITY AND INFORMATION HIGHWAYS

The following categorization of economic activities on information highways is only a proposal and may be somewhat of an oversimplification. But, for the purposes of discussion, the following two-tiered approach is taken despite some definitional overlap.

3.1. Type I: Networked Economic Activity

Type I economic activity may be defined as that activity which resides largely off the highway but makes use of the highway. These activities might be catalogued as:

- transshipment (information exchange)
- transaction (money as information exchange)
- transposed business function (making use of the networks to be more efficient and productive in current activities, direct client links, etc.)

Type 1 activities can be seen in a simple general model for business activity which makes use of networks:

seller		buyer
info exchanger A		info exchange B
transmitter	<---> technical network <--->	receiver
transaction initiator		transaction closer
info provider		info consumer

Policies for Type I activities may revolve around the use of highways in much the same way that general use policies (rules and regulations) are implemented for cars and trucks on physical transportation highways. The primary concern will be toward establishing equitable rules and regulations of the business activity itself. In other words, do businesses need to pass a "driver's test" to use the information highways?

3.2. Type II: Network Economic Activity

Type II economic activity may be defined as that which resides largely on the network itself and/or exists primarily because of the network environment. These activities might be catalogued as:

- network resident
- network "raison d'être"
- transformed business activity (activities which make use of networks to do new things)

Nearly all of Type II activities will inevitably be new business activities. (The somewhat lame transportation analogy is that of truck stops which is a service business that exists primarily because of the highway.) Many "non-economic" Type II network activities are already commonplace in the Internet environment including collaborative research and computer-aided cooperative work. With many businesses setting up "pages" on the World Wide Web of the Internet, we are beginning see the transformation of some economic activity to the network environment although most of this, for now, is still largely Type I activity. When this will change is when the primary business function exists and operates because of the networks.

Type II economic activities can readily provide for a number of challenges to conventional business wisdom. To see this, an example from broadcasting might be appropriate. In current networked designs (Type I), broadcasters are largely paid by advertisers to send the advertiser's message to the viewer. That is, advertisers pay for viewers indirectly through air time (i.e., channel space). The viewer, to the extent that they pay to receive the programming, is also indirectly paying to receive the advertising. Or, in the free-to-air case, the "free" programs come at the cost to viewers of more or less involuntary reception of the advertising as well.

However, in a network activity environment (Type II), such a system can easily be turned on its head. That is, the new wisdom may be for advertisers to pay viewers to watch ads since channel space or air time is no longer an issue. It would probably be done indirectly through levels of content provision such that an "all-advertising" information exchange would be free and "all-content" would be priced highest. (This is already being done indirectly with free magazine subscriptions to readers of many trade journals which are packed full of ads and the very high subscription costs for timely trade information newsletters which are largely without ads.) On information highways, "the truth may be free" which is probably why the user should be paid to receive advertising.

Two-way media are, by definition, much more interactive and content-focused. Therefore, broadcasting on the information highway network as a different activity made possible by the network (Type II activity) may, in turn, challenge the model of advertising and therefore the revenue base for such activity. The point is, as a policy issue, pricing of services will need to be reexamined, especially those that have had direct and indirect links to regulated public service tariffs in the past. Ownership issues and service licensing will become largely irrelevant in terms of broadcast service areas as the networks are geographically unbounded. What may arise instead is a need to focus policy toward a service standard and to be watchful for activity which may attempt to overly-consolidate (monopolize) service provision (a service anti-trust policy?) Certainly, strong legislation and enforcement of anti-criminal business activity will be required especially in the realm of fraud to which the networks will be susceptible.

4.   A NEW BUSINESS MODEL?

Economic activity on information highways (Type I and II) may require a new model for the emerging business environment. This model might be seen

314

as two concurrent models of hyper-card stacks of both the physical and
virtual structure of the environment.

A. Physical structure: (business links)

Data/information/entertainment bases

Network links/switching (highways)

Interfaces

User terminals

B. Virtual structure: (business function)

<------------> **Providers:** information/data/entertainment "raw
                              materials"
                access providers

    <------------> **Consolidators:** information content packaging
                                       transaction handlers

Info-

    <------------> **Value-adders:** direct marketers
Highways   interfacers
                                        knowledge brokers
    <------------> **Users:**   clients
                                        consumers
                                        customers
                                        curious

Some businesses are currently better placed than others to function at
one of the three top levels. Some new business players will function even
better. What will succeed? (Well, how good is your crystal ball?) A necessary

if not sufficient ingredient includes the primary focus on user needs that the model implies. These involve from the users perspective; (1) usefulness, (2) cost effectiveness, (3) performance availability, and (4) ease of use (Inmon 1986).

In fact, it is likely that any form of information highway will not gain wide acceptance unless the development strategy of providing for users needs is adopted.

In any case, the network business environment and the development of new economic activities as different from current activities will necessarily lead to a different set of policy issues.

5.   TOWARD A NEW POLICY FOCUS

Having economic activity reside on the network itself as well as continued activity which makes use of the network means dealing at a policy level with issues of those activities which are different from the current focus. The catalog of issues on first glance may, in fact, not seem so different after all. What is new perhaps will be to focus more attention on them as a comprehensive set of issues (not to be dealt with in isolation) in light of the differences that network environments may provide.

"New" policy issues: back to the future?

A.   Protection against fraud (at all levels in the model above)

Are current policies and laws up to the challenge of new

networks? Must we wait for court action to provide policy

direction? The potential for fraudulent activity in network

environments might be the single most important criminal issue

to be tackled and the greatest stumbling block to the

development of businesses on the networks.

B.   Consolidator anti-trust which refers to policies surrounding

ownership and control of network service activity and the

monopolistic control of those services.

What scope of access will be provided to vested economic

interests? How should we deal with highway conduit providers

who are themselves content providers?

C.   Intrusion by value-adders

Perhaps a new twist on privacy issues: Should we require

permission from users and/or should we protect users from

intrusive network behavior? What redress will users have for

abusive behavior directed at them?

D.   Competition and open markets

What levels of competition should be set in order to develop

optimal levels of network expansion for business use?

To what extent should policies be geared toward providing

free-market access to network environments? This is linked to

network ownership itself and anti-trust. In many developing

countries, protection of business concerns on the network,

especially government-owned and/or -controlled may provide a

vigorous counterpoint. At the same time, many developing

countries will want an assurance of free access to markets

including nondiscriminatory access through information

highways.

E.   Universal access

Businesses on the information highway will be largely service

businesses. As with all service providers, they can offer wares

without necessarily providing a means for their utilization and often without a license. This means the notion of universal service must evolve into a notion of universal access as service providers will not normally provide the means of access. This will be especially so where private enterprise is encouraged to build the information highways.

F.   Trade in services

The World Trade Organization (WTO, formally the GATT) and the General Agreement on Trade in Services (GATS) will need to evolve a set of principles for the conduct of trade in services in the network environment. Sovereignty and changes in international relations brought on by international trade in services on information highways will no doubt intrude into the international policy debate. (One likely candidate to take a lead in this debate is Singapore.) Will there be a need for new national and international codes of conduct? How can they be enforced? Should protectionist measures such as discriminatory pricing of transmission services, local content laws, and mandated use of government networks be avoided? Can cross-border network services be proscribed in favor of local provision regulation? Should a proportionate service ethic evolve?

G.   Tariff structures

With new models of "who pays for what," a new twist to public

utility pricing will ensue as well as changes to accounting rates of

communication exchange across sovereign borders. If business

largely pays for the technical network linkages and

simultaneously has a vested interest in part of the content in

some countries while governments pay for infrastructure in

others, how will they be priced? Who will regulate this pricing?

Is the cost-of-use pricing mechanisms for the network itself

enough especially in areas of disproportionate accessibility? Will

market pricing best determine service rates or will new domains

of public services require special protection and subsidies?

H.   Copyright and intellectual property rights

Do copyrights adequately protect intellectual property rights in

the network environments? And what becomes of the principle of

"fair use"?

I.   Author's rights

Are new policies needed to protect author's rights (creative work

paternity, rights of association, and right to publish and

withdrawal from publication, for example), rights of integrity, and

other "moral" rights? These are also problems of author

protection in network environments where the principle of

national treatment no longer applies (Berne Convention).

J.   User power (groups and individuals; good and bad)

Could new types of unions evolve such as a union of

service-content creative developers, for example, which will press

for levels of work conditions and pay on behalf of members to all

network service providers? Will we need rules of conduct for

special interest/pressure groups? How should we deal with the

aberrant user/user group: sabotage, espionage, libel and

liability, etc.?

K.   Information control

What are the optimal levels of information control even as some

claim information is inherently uncontrollable in open network

environments? Despite the difficulties in practicing censorship in
an open network, many governments have been quick to learn
that gateway access and channel blocks provide effective, if not
complete, control to "unwanted" information such as
pornography and incendiary religious or political comments.

L.   International and local agencies

Is there a role for an international agency such as the
International Telecommunications Union (ITU)? In other words,
how do we establish policies and enforceable regulation on an
international level? Pekka Tarjanne (1994), Secretary-General of
the ITU, identified five principles as the cornerstone of
international regulation affecting the information highway: these
are internationalism, universalism, regulatory symmetry,
regulatory independence, and open access. No doubt questions
of policies on a global level will need to come to grips with these
not altogether compatible or easily definable principles. On a
local (national) level, is there a need to consolidate the cacophony
of regulatory and policymaking agencies which claim some
jurisdiction over matters relating to information highways?

This list is by no means complete nor does it indicate relative levels of priorities which will necessarily be different from place to place. What it does mean, however, is that the current policy debates which revolve around issues of technical standards, privacy, security, and structural universal service (rather than access) become relatively minor. What we might want to begin is a spirited discussion in all quarters of the policy issues that may really matter as more extensive Type I economic activities evolve and Type II economic activities inevitably develop as the information highways themselves develop.

REFERENCES

Bates, S., 1994, *The Potential Downside of the National Information Infrastructure*, monograph of the Annenberg Washington Program, Communication Policy Studies, Northwestern University, December.

Cate, F. H., 1994, "The National Information Infrastructure: Policymaking and Policymakers", *Stanford Law and Policy Review*, vol. 6, no. 1.

Gilbert, J., 1994, "Does the Highway Go South?", *Intermedia*, vol. 22, no. 5 (October/November).

Harris, L. E., 1995, "Moral Rights on the Information Superhighway", *Intermedia*, vol. 23, no. 3 (February/March).

Inmon, William H., 1986, *Technomics: The Economics of Technology and the Computer Industry*, Dow Jones Irwin, Homewood, Illinois.

Jussawalla, Meheroo, 1993, *Global Telecommunications Policies: The Challenge of Change*, Greenwood Press, Westport and London.

McQuaide, T., 1995, "What is the Role of the Umpire in the Digital Playground of Giants", *Intermedia*, vol. 23, no. 1 (February/March).

324

Mulgan, G. J., 1991, *Communication and Control*, Guilford Press, New York and London.

Tarjanne, Pekka, 1994, *Regulating the International Information Infrastructure*, Geneva, ITU.

ABOUT THE EDITOR AND CONTRIBUTORS

D. McL. Lamberton is a pioneer in Information Economics. He was a co-
founder of the Center for International Research in Communication and
Information Technologies in Melbourne, Australia. He is the editor of
Prometheus and serves on the editorial board of Information Economics and
Policy. He is currently a Visiting Fellow in the Urban Research Program at
the Research School of Social Sciences at the Australian National University
in Canberra.

Eli Noam is Director, Columbia Institute of Tele-Information Center,
Columbia University. He has served on the New York State Public Service
Commission and has published many books in particular on telecommu-
nications regulations and policies in Europe.

Mark M. Braunstein is Professor of Economics and Acting Dean of the School
of Library and Information Studies, University of California at Berkeley. He
has many publications on the economics of information and communications,
intellectual property rights, and technical compatibility of standards.

Dan J. Wedemeyer is Professor of Communications at the University of Hawaii and has chaired that department for six years. He serves as Secretary of the Pacific Telecommunications Council in Hawaii and edits all of the Council's conference volumes.

Syed A. Rahim has been in the Communications Program of the East-West Center for more than two decades. He has directed several projects on Communications Policy and Planning in the Asia-Pacific region and his research has been published. Currently he is researching the cultural effects of the Internet.

Anthony Pennings is an Associate Professor of Communications at the Victoria University in Wellington, New Zealand. Pennings has been active in organizing meetings of the Pacific Telecommunications Council and has co-edited some of the Council's conference volumes.

John Locke is a doctoral candidate in the interdisciplinary program of the University of Hawaii covering Communications Information Science and Library Studies. He is a futurist scholar with expertise in computer software.

Mark Hukill is Deputy Chairman of the School of Communications and Mass Media at Singapore's Nanyang Technology University. Hukill is co-chairman of the Program Committee and Research Awards Committee of the Pacific

Telecommunications Council. His publications cover telecommunication

investments and markets in the ASEAN region.

Editor

Meheroo Jussawalla is a Senior Fellow Emeritus at the East-West Center.

She has specialized in the field of information and telecommunications

economics. Her research has focused on a wide range of major issues

including telecommunications trade between Japan and the United States,

satellite orbital competition, the economics of intellectual property rights, and

privatization of monopolies in the Asia-Pacific region. She is currently

researching infrastructure growth in China. She has authored over a dozen

books.

INDEX

Access

    to geostationary orbits, 291-98

    to information, democratization of, 186

    to orbitals, 291-98

    to plain old telephones (POTs), 3

    to telecommunications services, 109, 222-26, 263

        subsidy to assure, 202

    universal, new policy issues in, 317-18

Accumulated experience, role of, in a global economy, 49

Adaptability, limits to, and a global economy, 51

Advanced Broadcast Video Services (ABVS), defined, 267

Advanced Intelligent Networks, 177

Advantage, comparative, 50-51

Advertising, on Internet, 204

Affinity, as the basis for telecommunications relationships, 90

Allocation, of telecommunications services, 222-26

Alphabetization, and exchange of knowledge, 123

Alphanumerical calculation of electronic signals, 126

Alternet, backbone for Internet services, 199

American Economic Association, 49

American Standard Code of Information Interchange (ASCII), 129

Extended Binary Coded Decimal Interchange Code (EBCDIC), 129

Farrell, Joseph, 152, 170

Faulhaber, G. R., 201

FAX networking, 72

FAX transmission, standardization of, 150-51

Federal Communication Commission, 10, 103

    changes in regulations of, 103

Federal Ministry of Posts and Telecommunications (Federal Republic

 of Germany), 102

Feedback, formalized unit of, 126

Fiberoptic telecommunication, 211-12

    eliminating with cellular mobile technology, 203

    installation by long-distance carriers, 191-92, 232, 245

    submarine, 285, 298

Finlay, M., 127

Firewalls, in Internet, 195

Fixed cost, of Internet, 200

Flannagan, Patrick, 283

Flexibility, digital versus analog devices, 218-22

Follow-me number, 71-72

Fordism, 134-35

Foucault, Michel, 110n.1, 130

Frame Relay Forum, 170

France, telecommunication culture of, 102